Ahmed Al-Mukhtar

Fracture mechanics method of welded components under cyclic loads

Ahmed Al-Mukhtar

Fracture mechanics method of welded components under cyclic loads

Fatigue life calculations and simulation

Südwestdeutscher Verlag für Hochschulschriften

Imprint
Any brand names and product names mentioned in this book are subject to trademark, brand or patent protection and are trademarks or registered trademarks of their respective holders. The use of brand names, product names, common names, trade names, product descriptions etc. even without a particular marking in this work is in no way to be construed to mean that such names may be regarded as unrestricted in respect of trademark and brand protection legislation and could thus be used by anyone.

Cover image: www.ingimage.com

Publisher:
Südwestdeutscher Verlag für Hochschulschriften
is a trademark of
Dodo Books Indian Ocean Ltd., member of the OmniScriptum S.R.L Publishing group
str. A.Russo 15, of. 61, Chisinau-2068, Republic of Moldova Europe
Printed at: see last page
ISBN: 978-3-8381-2301-1

Zugl. / Approved by: Freiberg, TU, Diss., 2010

Copyright © Ahmed Al-Mukhtar
Copyright © 2011 Dodo Books Indian Ocean Ltd., member of the OmniScriptum S.R.L Publishing group

To Dania and Bainat

ACKNOWLEDGEMENTS

I would like to acknowledge the *Free State of Saxony "Freistaates Sachsen"* for the award of scholarship that enabled me to undertake this research work at Technische Universität Bergakademie Freiberg, Institute of Materials Engineering, Germany.

My special thanks go to *Prof. Dr.-Ing. habil. H. Biermann*, Head of the Institute of Materials Engineering (TUBA Freiberg), *Prof. Dr-Ing. P. Hübner*, Faculty of Mechanical Engineering, University of Applied Science, Mittweida, and *Prof. Dr.-Ing. Uwe Zerbst* from Federal Institute for Materials Research and Testing (BAM) for many fruitful discussions and comments.

My thanks go to my colleague *Dr.-Ing. S. Henkel*, Institute of Materials Engineering (TUBA Freiberg), for his guidance, and technical support.

PREFACE

Fracture Mechanics process of Welded Joint is a very vast research area and has many possibilities for solution and prediction. Although the fatigue strength (FAT) and stress intensity factor (SIF) solutions are reported in several handbooks and recommendations, these values are available only for a small number of specimens, components, loading and welding geometries. The available solutions are not always adequate for particular engineering applications. Moreover, the reliable solutions of SIF are still difficult to find in spite of several SIF handbooks have been published regarding the nominal applied SIF. The effect of residual stresses is still the most challenge in fatigue life estimation. The reason is that the stress distributions and SIF modified by the residual stresses have to be estimated. The stress distribution is governed by many parameters such as the materials type, joint geometry and welding processes.

In this work, the linear elastic fracture mechanics (LEFM), which used crack tip SIFs for cases involving the effect of weld geometry, is used to calculate the crack growth life for some different notch cases.

The variety of crack configurations and the complexity of stress fields occurring in engineering components require more versatile tools for calculating SIFs than available in handbook's solutions that were obtained for a range of specific geometries and load combinations.

Therefore, the finite element method (FEM) has been used to calculate SIFs of cracks subjected to stress fields. LEFM is encoded in the FEM software, FRANC, which stands for fracture analysis code.

The SIFs due to residual stress are calculated in this work using the weight function method.

The fatigue strength (FAT) of load-carrying and non-load carrying welded joints with lack of penetration (LOP) and toe crack, respectively, are determined using the LEFM. In some studied cases, the geometry, material properties and loading conditions of the joints are identical to those of specimens for which experimental results of fatigue life and SIF were available in literature so that the FEM model could be validated.

For a given welded material and set of test conditions, the crack growth behavior is described by the relationship between cyclic crack growth rate, da/dN, and range of the stress intensity factor (ΔK), i.e., by Paris' law. Numerical integration of the Paris' equation is carried out by a FORTRAN computer routine. The obtained results can be used for calculating FAT values. The computed SIFs along with the Paris' law are used to predict the crack propagation. The typical crack lengths for each joint geometry are determined using the built language program by backward calculations.

To incorporate the effect of residual stresses, the fatigue crack growth equations which are sensitive to stress ratio R are recommended to be used. The Forman, Newman and de Konig (FNK) solution is considered to be the most suitable one for the present purpose.

In spite of the recent considerable progress in fracture mechanics theories and applications, there seems to be no, at least to the author's knowledge, systematic study of the effect of welding geometries and residual stresses upon fatigue crack propagation based completely on an analytical approach where the SIF due to external applied load (K_{app}) is calculated using FEM. In contrast, the SIF due to residual stresses (K_{res}) is calculated using the analytical weight function method and residual stress distribution. To assess the influence of the residual stresses on the failure of a weldment, their distribution must be known.

Although residual stresses in welded structures and components have long been known to have an effect on the components fatigue performance, access to reliable, spatially accurate residual stress field data are limited. This work constitutes a systematic research program regarding the concept for the safety analysis of welded components with fracture mechanics methods, to clarify the effect of welding residual stresses upon fatigue crack propagation.

Ahmed Al-Mukhtar

EINLEITUNG

Die Bewertung einer Schweißnaht ist ein großes Forschungsgebiet und hat viele Möglichkeiten für Lösungskonzepte und Vorhersagen. Obwohl für die Schwingfestigkeit und die Spannungsintensitätsfaktor (SIF)-Lösungen in verschiedenen Handbüchern Empfehlungen ausgewiesen sind, sind diese Werte nur für eine geringe Anzahl von Proben, Komponenten, Belastungsfälle und Schweißgeometrien verfügbar. Die vorhandenen Lösungsansätze sind nicht immer für spezielle technische Anwendungen geeignet. Darüber hinaus sind zuverlässige bewährte Lösungen von Spannungsintensitätsfaktoren immer noch schwierig zu finden, obwohl verschiedene SIF-Handbücher mit Hinweis auf den anliegenden nominalen SIF veröffentlicht sind. Der Einfluss von Eigenspannungen ist eine der größten Herausforderungen bei der Lebensdauerabschätzung. Aufgrund der Tatsache, dass infolge der Eigenspannungen sowohl die Spannungsverteilung als auch der SIF verändert werden, muss eine Abschätzung erfolgen. Die Spannungsverteilung wird durch viele Parameter beeinflusst, wie zum Beispiel den Werkstoff, die Nahtgeometrie und den Schweißprozess.

In der vorliegenden Arbeit wurde für die Berechnung des Ermüdungsrisswachstums unter verschiedenen Kerbfällen das Konzept der linear-elastischen Bruchmechanik (LEBM) verwendet, welches K-Lösungen für die Rissspitze bei unterschiedlichen Fällen der Schweißgeometrie berücksichtigt.

Aufgrund der Komplexität der Risskonfigurationen und der Spannungsfelder in praxisrelevanten Komponenten werden weitere Hilfsmittel zur Berechnung von Spannungsintensitätsfaktoren benötigt, welche die herkömmlichen Lösungen in Handbüchern erweitern.

Deshalb wurde die Finite Elemente Methode (FEM) zur Berechnung von Spannungsintensitätsfaktoren an Rissen verwendet. Die LEBM wird in der FEM-Software FRANC berücksichtigt.

Die aus Eigenspannungen resultierenden Spannungsintensitätsfaktoren wurden mit Hilfe der Gewichtsfunktionsmethode berechnet.

Die Ermüdungslebensdauer (Schwingfestigkeit) von tragenden und nichttragenden Schweißnähten mit ungenügender Durchschweißung beziehungsweise Kerbriss wurden mit Hilfe der LEBM durch Integration der Zyklischen Risswachstumskurve ermittelt. Zur Validierung des FEM-Modells konnte in einigen untersuchten Fällen auf experimentelle Ergebnisse zur Lebensdauer und zum SIF aus der Literatur zurückgegriffen werden, wo identische Geometrien, Materialeigenschaften und Belastungsverhältnisse der Naht vorlagen.

Unter Vorgabe des Werkstoffes und der Prüfbedingungen wurde das Risswachstumsverhalten mit dem Zusammenhang von Risswachstumsgeschwindigkeit da/dN und zyklischem Spannungsintensitätsfaktor ΔK mit dem Paris-Gesetz beschrieben. Eine numerische Integration der Paris-Gleichung erfolgte über ein FORTRAN-Programm. Die damit erhaltenen Ergebnisse sind als Ermüdungslebensdauer (Schwingfestigkeit) verwendbar. Die berechneten SIF'en entlang der Paris-Geraden werden zur Vorhersage des Risswachstums benutzt. Die typischen Risslängen für jede Nahtgeometrie wurden mit Hilfe des eigens integrierten Programmes ermittelt.

Zur Berücksichtigung des Einflusses von Eigenspannungen wird empfohlen, Risswachstumsgleichungen zu nutzen, die empfindlich auf das Spannungsverhältnis R reagieren. Für die vorliegende Zielsetzung gilt der Lösungsansatz nach Forman, Newman und de Konig (FNK) als der am besten geeignete.

Trotz der jüngsten, beträchtlichen Fortschritte in den bruchmechanischen Theorien und Anwendungen sind systematische Studien zum Einfluss der Schweißgeometrie und der Eigenspannungen auf das Ermüdungsrisswachstum, in welchen der SIF aufgrund extern anliegender Beanspruchungen (K_{app}) mit der FEM berechnet wurde, in der Literatur kaum vorhanden. Im Gegensatz dazu wurde der SIF infolge von Eigenspannungen (K_{res}) mit Hilfe der analytischen Gewichtfunktionsmethode und der Eigenspannungsverteilung berechnet. Um den Einfluss von Eigenspannungen auf das Versagen einer Schweißverbindung abzuschätzen, muss deren Verteilung bekannt sein.

Obwohl die Wirkung von Eigenspannungen auf das Ermüdungsverhalten in geschweißten Strukturen und Komponenten schon lange bekannt ist, ist der Zugriff auf verlässliche und präzise Daten von räumlichen Eigenspannungsfeldern begrenzt. Bezüglich einer konzeptionellen Sicherheitsanalyse von geschweißten Komponenten mit bruchmechanischen Methoden begründet diese Arbeit einen systematischen Ansatz, um den Einfluss von Schweißeigenspannungen auf das Ermüdungsrisswachstum zu verdeutlichen.

Ahmed Al-Mukhtar

TABLE OF CONTENTS

ACKNOWLEDGEMENTS 3
PREFACE... 5
EINLEITUNG 7
TABLE OF CONTENTS.. 11
NOMENCLATURE AND ABBREVIATIONS 15

Chapter One 19
INTRODUCTION 19
1.1. Background and Motivation 19
1.2. Fatigue Failure 20
1.3. Failure of Welded Joints 23
1.3.1. Butt Welded Joints 24
1.3.2. Cruciform Fillet Welded Joints 25
1.3.2.1. Non-Load Carrying Fillet Weld 26
1.3.2.2. Load-Carrying Fillet Weld 26
1.4. Concept of Fracture Mechanics Methods 27
1.5. Residual Stresses in Welding 28
1.6. Objectives and Scope 29
1.7. Outline 30
1.8. The Present Work 31

Chapter Two 33
OVERVIEW OF FAILURE OF WELDED JOINTS 33
2.1. Introduction 33
2.2. Fracture Mechanics of Fillet Welded Joints 33
2.3. Crack Propagation Curve.. 40
2.4. Simulation of Fatigue Crack Growth 43
2.5. Fracture Analysis Code 46
2.6. Factors Affecting the Fatigue Strength of Welded Joints 47
2.6.1. Weld Geometry... 47
2.6.2. Weld Defects and Metallurgy 55
2.6.3. Materials and Welding Techniques 56
2.6.4. Weld Residual Stresses 56
2.7. Residual Stress Effects on Fatigue Life of Welded Joints....................... 57
2.7.1. Crack Closure and Effective SIF 59
2.7.2. Determination of Residual Stresses 61
2.8. Superposition Method 64
2.8.1. Residual Stresses Distribution 65
2.8.1.1. R6 Distributions for T-Plate 66
2.8.1.2. Polynomial Distributions for T-Plate and Butt Welds.......................... 66

2.8.2. Residual Stress Intensity Factor 68
2.9. Conclusions and Discussion for the Current Work.. 74

Chapter Three.. 78
MODELING OF WELDED JOINTS AND SIF CALCULATION................................... 78
3.1. Introduction 78
3.2. Two Dimensional Analysis of Welded Joints... 79
3.3. Finite Element Analysis.. 81
3.4. Mesh Description and Boundary Conditions 82
3.5. Material Properties.. 83
3.6. Solution Procedure 84
3.6.1. Mesh Generation 85
3.6.2. Selection of Material Model... 86
3.6.3. Crack Propagation 86
3.6.4. Modeling Procedures 87
3.7. Results Convergence.. 89
3.7.1. Influence of Mesh Size... 89
3.7.2. Influence of Crack Increment 91
3.7.3. Influence of Mesh Type.. 93
3.7.4. Influence of Symmetry 95
3.8. Selection of the Notch Cases... 98
3.9. Stress Intensity Factor Calculations... 99
3.9.1. SIF of Load-Carrying Cruciform Joints... 100
3.9.2. SIF of Non-Load-Carrying Cruciform Joints 103
3.9.3. SIF of Transverse Butt Weld Joints having Toe Crack.................................. 105
3.9.4. SIF of Transverse Butt Weld joints having LOP 106
3.10. Results and Discussion... 107
3.10.1. Cruciform Fillet Weld Model 107
3.10.2. Stress Distribution... 107
3.10.3. Butt Weld Model.. 112
3.10.4. Results of Verifications 113
3.10.4.1. SIF of Non-Load-Carrying Cruciform Joints 113
3.10.4.2. SIF of Load-Carrying Cruciform Joints.. 115
3.10.4.3. SIF of Transverse Butt Weld Joints having Toe Crack............................. 117
3.10.4.4. SIF of Transverse Butt Weld Joints having LOP 117
3.11. Effect of Geometry in Load-Carrying Cruciform Joints.................................. 118
3.11.1. Effect of Weld Shape 119
3.11.2. Effect of Plate Thickness Ratio 120
3.12. Conclusions 122

Chapter Four.. 124
FATIGUE LIFE CALCULATIONS AND VERIFICATIONS 124
4.1. Introduction 124

4.2. Fracture Mechanics Analysis 126
4.3. Fatigue Life Calculations Using Paris' Law 128
4.4. Fatigue Life Calculation Using NASGRO Equation 131
4.5. Parameters of Crack Propagation Life 133
4.6. Selection of the Notch Cases 134
4.7. Backward Calculation of Fatigue Strength 135
4.8. Results and Discussion 138
4.8.1. Fatigue life Calculations 138
4.8.1.1. Cruciform Weld Joints having Toe Crack 139
4.8.1.2. Transverse Butt Weld Joints having Toe Crack 143
4.8.1.3. Cruciform Weld Joints having LOP 144
4.8.1.4. Transverse Butt Weld Joints having LOP 147
4.9. Experimental Verifications 151
4.10. Effect of Residual Stresses 160
4.11. Effect of Sheet Thickness 160
4.12. Geometrical Verification of Butt Weld Joints 168
4.12.1. Simulation of Fatigue Crack Growth 169
4.12.2. Failure Mode 170
4.12.3. Weld Metallurgy and Defects 171
4.12.4. Fatigue Life and Crack Growth 172
4.12.5. Modeling 174
4.12.6. Verification Results 174
4.12.7. Standards Verifications 179
4.13. Conclusions 181

Chapter Five 183
RESIDUAL STRESSES IN WELDED JOINTS 183
5.1. Introduction 183
5.2. Assumptions in Finite Element Model 184
5.3. Residual Stresses Assessment 185
5.4. Residual Stress Intensity Factor 187
5.5. Residual Stresses Distribution 191
5.6. BSI 7910 Distribution 193
5.7. Total Stress Intensity Factor 194
5.8. Crack Growth Propagation Life 196
5.9. The FNK Equation 196
5.10. Results Verification and Procedures 197
5.11. Effect of Residual Stresses Distribution on SIF 200
5.12. Predicting Fatigue Crack Growth Rates 202
5.13. Effect of Residual Stresses and Stress Ratio on FCG Rate 205
5.14. S-N Curve 213
5.15. Conclusions 216

Chapter Six 218
CONCLUSIONS AND RECOMMENDATIONS FOR FUTURE WORK 218
6.1. Conclusions 218
6.2. The Originality of the Current Calculations 221
6.3. Recommendations for Future Work 222

REFERENCES 224

NOMENCLATURE AND ABBREVIATIONS

Notations	Meaning
a	Crack length.
a_i	Initial crack length.
a_f	Final crack length.
A_0, A_1 and A_2	Polynomial function of weld size.
$A_{0,n}$, $A_{1,n}$, $A_{2,n}$, and $A_{3,n}$	Polynomial coefficient in Newman's equation.
BEM	Boundary element method.
B	Attached plate thickness (Cruciform joint).
BSI	British Standard Institution.
C	Material depend constant (Paris' law).
C_{mean}	The mean fatigue crack growth rate coefficient corresponding to 50% survival probability value.
C_{char}	The characteristic value corresponding to 95% survival probability value.
C_{ESA}	Materials depend constant (NASGRO Equation).
da/dN	Crack growth rate.
E	Young's modulus.
FEM	Finite element method.
FNK	Forman, Newman and de Konig.
f	Newman's effective stress ratio.
FAT_{mean}, $(FAT95\%)$	The fatigue strength corresponding to 50% survival probability.
FAT_{char}, $(FAT50\%)$	The fatigue strength corresponding to 95% survival probability.
$f(a/t)$	Geometrical function (crack length/thickness).
$f(t)$	Reduction factor.
FAT	Fatigue strength at two million cycles (FAT95%).
FCG	Fatigue crack growth.
h	Fillet welds leg length on main plate side (Cruciform joint).
H	Weld bead height (Butt weld).
HAZ	Heat-affected zone.
IIW	International Institute of Welding.
K	Linear elastic stress intensity factor (SIF).
ΔK_{app}	Applied stress intensity factor.
ΔK	Stress intensity factor range ($\Delta K = K_{max} - K_{min}$).
K_{max} or $K_{max,app}$	Stress intensity factor due to maximum applied load.
K_{min} or $K_{min,app}$	Stress intensity factor due to minimum applied load.
K_t	Theoretical stress concentration factor.
K_{IC}	Fracture toughness.
K_I	Stress intensity factor mode-I.

K_{II}	Stress intensity factor mode-II.
K_c	Critical stress intensity factor.
K_{th}	Threshold stress intensity factor.
K_{eff}	Effective stress intensity factor.
K_{op}	Crack opening stress intensity factor.
K_{res}	Stress intensity factor introduced by residual stresses field.
K_T	Total stress intensity factor ($K_T = K_{app} + K_{res}$).
k	Material constant.
q	Arc welding power.
v	Weld travel speed.
η	Welding process efficiency.
LOP	Lack of penetration.
L	Plate length.
LEFM	Linear elastic fracture mechanics.
$m(x,a)$	Weight function.
m	Material depends constant (Paris' law).
n	Material depends constant (NASGRO Equation).
n_{red}	Reduction factor exponent.
N	Number of cycles.
N_p	Propagation number of cycles.
N_I	Initial number of cycles.
N_T	Total number of cycles ($N_T = N_I + N_p$).
r_o	Size of plastic zone.
R	Stress ratio ($\sigma_{min}/\sigma_{max}$).
R_{app}	Applied stress ratio ($\sigma_{min}/\sigma_{max}$).
R_{res}	Residual stress ratio ($\sigma_{min}+\sigma_{res})/(\sigma_{max}+\sigma_{res}$).
R_{eff}	Effective stress ratio ($K_{max}-K_{open})/(K_{max}-K_{min}$).
R'	Crack opening ratio ($\sigma_{max}-\sigma_{open})/(\sigma_{max}-\sigma_{min}$).
S	Fillet welds leg length on attached plate side (Cruciform joint).
STDV	Standard deviation.
S_o	Material flow stress.
SIF	Stress intensity factor.
S-N curve	Fatigue design curve (Wöhler curve).
S_{max}	Peak stress.
t	Plate thickness.
t_o, t_{eff}	Reference thickness used in reduction factor.
T	Main plate thickness (Cruciform joint).
WRS	Welding residual stress.
W	Weld bead width (Butt weld).
w	Fillet welds width in cruciform joint ($w=B+h$).
x	Vector direction from the edge of crack surface to a for each Δa.

ρ	Weld toe radius.
θ	Weld toe angle ($\gamma=180-\theta$).
ε	Strain.
$\sigma,$	Stress.
σ_n	Nominal stress.
σ_y	Yield strength (same as S_{ys}).
α	Material constant (NASGRO Equation).
υ	Poisson's ratio.
σ_{max}	Maximum stress.
σ_{min}	Minimum stress.
σ_{open}	Crack opening stress.
$\sigma(x)$	Stress distributions.
$\Delta\sigma$	Stress range.
σ_{cr}	Critical stress.
$\Delta\sigma_e$	Endurance stress range.
σ_{res}	Residual stress distribution.
2-D	Two dimensional models.
3-D	Three dimensional models.
Exp.	Experimental (Experimental verification section).
Prop.	Propagation (Fatigue life calculation and comparison section).
SINT.	SINTAP profiles (Residual stress distribution).

Chapter One
INTRODUCTION

1.1. Background and Motivation

Engineering structures and components consist of single or several parts which connect each to other by joining processes. These parts are manufactured by several processes such as casting, forging, rolling and extrusion etc. The components can be fabricated by joining theses parts by riveting, bolting or welding the parts together.

Welding is today the most common joining method for metallic structures. Its industrial application is extremely important and many of the large structures designed and erected in the last decades would not have been possible without modern welding technology. Typical examples are steel bridges, ship structures, and large offshore structures for oil exploitation. The strength analysis of welded structures does not deviate much from that for other types of structures. Various failure mechanisms have to be avoided through appropriate design, choice of material, and structural dimensions. Design criteria such as yielding, buckling, creep, corrosion, and fatigue must be carefully checked for specific loading conditions and environments. It is, however, a fact that welded joints are particularly vulnerable to fatigue damage when subjected to repetitive loading. Fatigue cracks may initiate and grow in the vicinity of the welds during service life even if the dynamic stresses are modest and well below the yield limit. The problem becomes very pronounced if the structure is optimized by the choice of high strength steel. The very reason for this choice is to allow for higher stresses and reduced dimensions, taking benefits of the high strength material with respect to the yield criterion. However, the fatigue strength of a welded joint is not primarily governed by the strength of the base material of the joining members; the governing parameters are mainly the global and local geometry of the joint. Hence, the yield stress is increased, but the fatigue strength does not improve significantly [1]. This makes the fatigue criterion a major issue. The fatigue strength will alone give the requirements for the final dimensions of the structural members such as plates and stiffeners as well as welding processes and geometries.

Welding processes lead to deformation and creation of local and global residual stress fields. Moreover, they are the source of crack-like defects. Although of these disadvantages, welding is one of the major tools for production of components, machines and constructions, respectively. It is a cost saving and weight reducing process.

Weld crack-like defects such as the weld toe and the weld root cracks are partially present in welded engineering structures. The ability to evaluate the effect of these cracks on fatigue life of the welded joints is necessary. To carry out this evaluation, linear elastic fracture mechanics (LEFM) is recommended and highly attractive.

There are special problems in the fracture mechanics assessments of welded components that are the lack of the information on the local notch action of the welding joints and the residual stresses that were induced during the welding. Therefore, up to date these assessments are based on conservative limits of crack lengths and in most cases, the residual stress distributions are not taken into account.

The assumption of high residual stress distributions produces mostly too short lives which are not in correlation with the real experimental results. In addition, the assumption of an incorrect crack length leads to serious misinterpretations.

Researches to date are based mainly on experimental tests to evaluate the fatigue properties, which mainly require high time and cost in case of testing lot of welded joints and materials. Little attention is paid on the analytical modelling of fatigue and fracture behaviour.

1.2. Fatigue Failure

Fatigue is defined as damage accumulation due to oscillating stresses and strains in the material. Therefore, fatigue cracks do occur in welded details that are subjected to repetitive loading. In significant structural items they may lead to failures with severe consequences [1].

Fatigue is the main cause of damage, followed by groups that can be designated as accidental damage. Figure 1.1 shows a fatigue failure of a propeller shaft in a shuttle tanker. The fracture occurred in the intermediate part of the shaft. The crack started from the surface of the shaft due to a weld arch strike. The fatigue surface is characterized by its smooth appearance with almost no plastic strain. At several stages during crack propagation, marks which are due to low stress variations are left as traces on the fatigue surface. These so-called beach marks correspond to changes in the fatigue loading; the crack front will make a mark during the time of slow growth due to smaller stress cycles. These marks are analogous to the dark winter rings found in the cross section of a tree. As can be seen, the beach marks have a typical semi-elliptical shape indicating the position of the crack front at various stages during the crack propagation. When the fatigue crack has reached the size of about three-quarters of the shaft diameter ($D = 360$ mm), the final fracture has occurred due to lack of the remaining ligament of the shaft cross section. It is a ductile fracture governed by the maximum occurring shear stress [1].

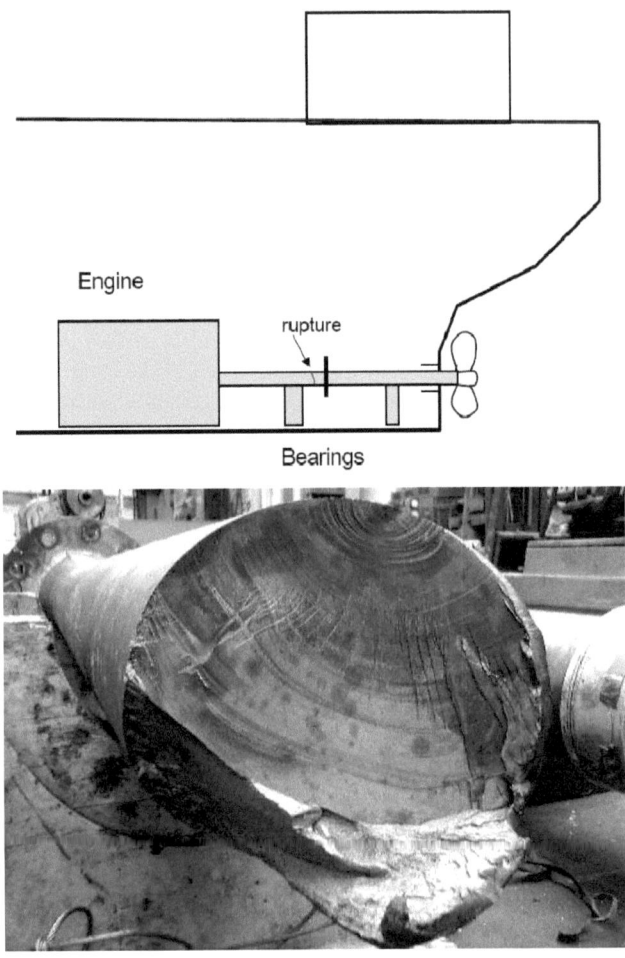

Figure 1.1: Fatigue failure in a propeller shaft in a shuttle tanker. Crack has initiated from a weld arch strike at the surface [1]

In welded structures, welds are the critical connections in fabrication structures because crack-like defects are more commonly present in the welds than in the parent material.

Moreover, material properties and stresses associated with welds are more variable than in parent material. Thus, failure is more likely to occur at welds. In 1968, a catastrophic failure occurred in a 350 MW turbine at the intermediate pressure loop-pipe to steam chest weldment in U.K. The failure was traced to the circumferential cracking at the interface between the 1 Cr-1 Mo-0.3 V cast steam chest and the 2 ¼ Cr-1 Mo weld metal. The crack was believed to have been introduced during the heat treatment operation or within a short time of commissioning of the turbine. A fracture mechanics approach was adopted to analyse the safety of the components and those found not suitable were replaced. Thus, application of fracture mechanics can be applied in the broad area of determining the acceptable defect's size.

In the same time the incidence of fatigue cracking has become a problem of critical importance. Welded joints have normally complicated geometries and are subjected to complicated service environments and stress conditions which may finally lead to early fatigue cracking.

Fatigue and fracture analysis of cracks are therefore of great practical interest and require the accurate determination of the stress state at a crack tip which is defined in terms of the stress intensity factor, SIF for the case under analysis. They are necessary for calculating the fatigue life.

1.3. Failure of Welded Joints

Welds can introduce severe stress concentrations which differ from one structural element to another [1, 2].

Figure 1.2 compares the fatigue performances of an E36 steel plate, and the same plate with a hole or two stiffeners welded longitudinally. Fatigue strength (FAT), which is conventionally given as 2 million cycles, passes respectively from 260 MPa (plain plate) and 180 MPa (bored plate) to 70 MPa for the welded joint [2]. To eliminate the possibility of rupture, a large safety coefficient can be applied. Then the welded structure becomes too large and is not competitive. Hence, the advantages of fracture mechanics will be taken.

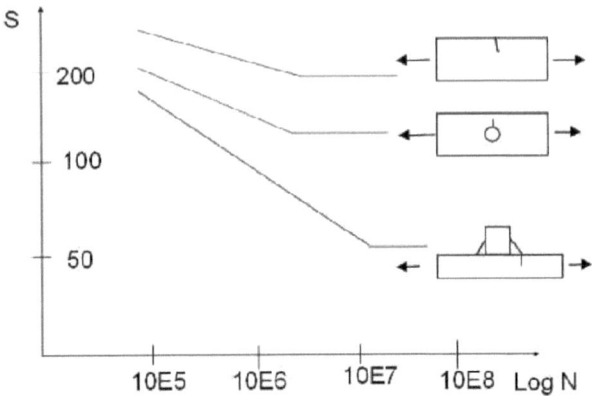

Figure 1.2: Fatigue life curves for various details [1, 2]

Welding operation result in the existence of stress concentrated areas where fatigue crack may originate and expand from. Welded joint geometries, internal defects (lack of penetration, LOP) or external defects (undercuts, slag inclusions), and loading types will determine the crack path (CP) and fatigue life of a specific joint. The most traditional welded joints are described below.

1.3.1. Butt Welded Joints

Figure 1.3 shows transverse butt welds. Two plates are joined via a transverse welding perpendicular to the load axis [2]. This joint is widely used in welded structures.

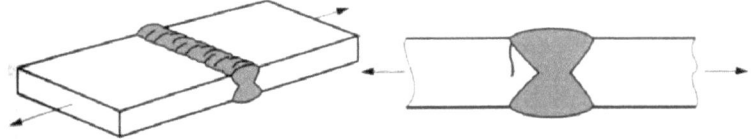

Figure 1.3: Fatigue cracking of a transverse butt weld [2]

For this type of joint, the fatigue crack starts at the weld toe and propagates through the thickness of the sheet, perpendicular to the load direction. The crack is thus not the result of a defect including welding or bad properties of the deposit metal, but the consequence of stress concentrations at the weld toe.

In this type of butt welded joint, the influence of the shape of the weld bead is important for determining the endurance characteristics of the joint. This depends greatly on the welding conditions.

Other types of butt joints are longitudinal butt welds. The load is parallel with the direction of the welding. The fatigue performance of these welds is better than that of transverse joints. However, this is not always the case. Therefore, this book presents a procedure for determining fatigue strength of different joints.

Figure 1.4 shows that the crack in this type of joints generally starts at the level of a welding stop and restart, for examples, while changing an electrode, or starting from a deformation on the weld bead surface. A good fatigue strength of longitudinal joints can only be obtained if they are continuous, and therefore if welding interruptions are avoided.

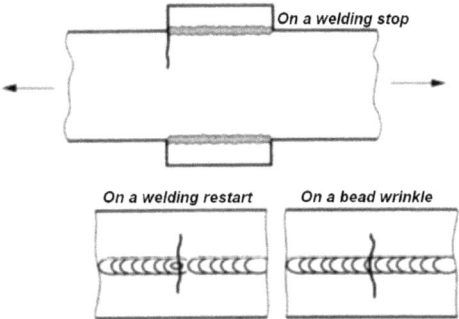

Figure 1.4: *Fatigue cracking of a longitudinal butt weld [2]*

1.3.2. Cruciform Fillet Welded Joints

There exist two types of cruciform joints, according to whether the weld beads transmit the load or not, i.e., load-carrying and non-load carrying cruciform joints, respectively.

1.3.2.1. Non-Load Carrying Fillet Weld

Figures 1.5(a) and 1.5(b) show the fillet weld which does not transmit the load in the solid metal sheet. In this case the crack starts at the weld toe and propagates through the thickness of the plate in a plane perpendicular to that of the applied stress [2, 3].

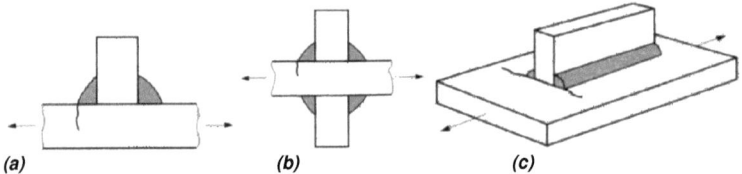

(a)   *(b)*   *(c)*

Figure 1.5: *Fatigue cracking of a non-load carrying fillet weld [2]*

There is no advantage in making assemblies with fillet welds parallel to the stress direction (see Figure 1.5c). The crack then starts at the bead end and leads to a low fatigue strength. On the other hand, continuous longitudinal fillet welds present significant improvements in endurance over intermittent fillet welds [2].

1.3.2.2. Load-Carrying Fillet Weld

In this type of joint, the entire load is transmitted by the weld (see Figure 1.6a). In addition to the stress concentration zones located at the weld toe there are zones of actually angled internal notches at the weld root. In general the crack starts there and then propagates in the deposited metal in an oblique direction compared to the load direction.

The fact of increasing the throat thickness, by a thicker weld or a better penetration, is not always enough to ensure that the fatigue crack starts, this time, at the weld toe. In general, completely interpenetrating bead (see Figure 1.6b) notably improve the endurance properties.

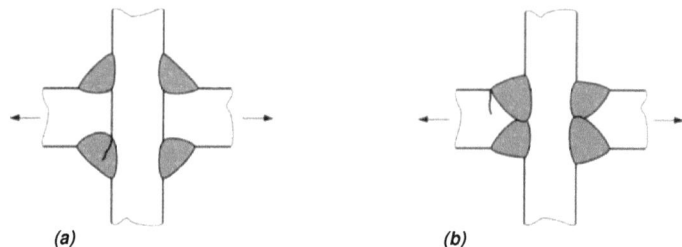

Figure 1.6: Fatigue cracking of a load-carrying fillet weld [2]: (a) root crack; (b) toe crack

1.4. Concept of Fracture Mechanics Methods

With the consideration of welding as the major method of fabrication, cracking in the heat-affected zone (HAZ) and weld metal has become a serious problem particularly in large and continuous structures. The inevitable cracking in weld joints makes the estimation of initial crack length (a_i) to be very important to calculate the time of fatigue crack growth propagation (FCG). In particular, the fact that fatigue cracks initiate very readily at the weld toe, virtually eliminating a crack initiation period and giving a fatigue life which is spent largely in crack propagation [4].

The weld cracks are induced due to various factors. For example, one of the major problems in the welding of steels is the type of cracking which is generally known as hydrogen induced cold cracking. At high carbon levels, the HAZ cracking was severe because of the formation of brittle martensite as a result of rapid cooling in welding. In the presence of hydrogen gross cracking is inevitable. The combination between the residual stress and corrosive media also produce a cracking called stress corrosion cracking.

Regardless the reasons for crack formations, fracture mechanics is mostly used in life prediction of a welded structure with an existing crack or under assumptions of the presence of a crack not found by non-destructive testing.

The initiation phase is assumed negligible for welded joints in the fracture mechanics approach and the life calculation is based upon a stress intensity factor SIF, which accounts for the magnitude of stress, crack size and joint details geometries.

Regarding the first crack nucleation phase, it is possible to use a local, cyclic stress strain approach at the weld notch to calculate the time to crack initiation. However, the same practical result can be obtained by a fracture mechanics approach using a fictitious, small, initial crack size [1].

It is to be interested that these small, initial cracks must be regarded as fictitious cracks chosen to make LEFM describe the entire fatigue process. They are not real, physical cracks (flaws, intrusions) created by the welding procedures. It is therefore difficult to relate the derived model to actual detected flaws in the weld. Furthermore, the initial cracks are often so small that LEFM is not applicable, typical values being from a_i= 0.01 to 0.05 mm [1], or even 0.15 which is recommended by IIW [3, 34]. This crack size is close to the microstructural features of the material. Ordinary structural steel qualities often have a grains size near 0.01 mm. Therefore, it is to be emphasized that it may be dubious to apply LEFM at crack depths of less than 0.1 mm. Before this stage the crack initiation phenomenon is probably better modeled by the Coffin and Manson equation. This equation is based on a local stress-strain approach [1]. The J-integral method and Paris' crack growth law have used also to predict crack growth rates and identical fatigue behaviour in short crack propagation. Nevertheless, fracture mechanics approach is used in this work, and the evaluation of the ability of this tool in fatigue life calculation is presented.

1.5. Residual Stresses in Welding

Residual stresses are internal forces in equilibrium themselves. These stresses are produced due to various manufacturing processes such as forging, casting, rolling, machining, cutting, heat treatment, surface treatment, hardening and shot peening. The formation of some microstructures such as bainite and martensite lead also to residual stresses. Moreover, residual stresses develop due to the welding processes which are concerned in the current work.

Residual stresses are important in fatigue life, stress corrosion resistance, dimensional stability and brittle fracture. In general, the effect of residual stresses may either be beneficial or detrimental, depending on magnitude, sign and distribution of the stresses with respect to the load-induced stresses. Tensile residual stresses are detrimental and often in the magnitude of the materials yield strength. The tensile residual stresses will reduce the fatigue life of the structure by increasing the growth of a fatigue crack, while compressive residual stresses are very beneficial and will decrease the fatigue crack growth rate [5-9].

In welding the residual stresses are developed due to non-uniform thermal expansion caused by the local heating of the structure. The yield stress is strongly temperature dependent, so the maximum stress at any point in the metal depends on the local temperature. The temperature will vary in three dimensions as the weld progresses, through the thickness, across the width of the weld and along the length. This gives rise to a complex stress distribution throughout the weldment which will be further complicated as a weld subsequently is deposited.

Residual stresses will never contribute to failure by plastic collapse but they will make a significant contribution to failure by brittle fracture (linear elastic fracture), or stress corrosion cracking in susceptible environments. Thus to assess the influence of the residual stresses on the failure of a weldment, their distribution must be known.

It is duly expected that in a notched specimen not only the phenomenon of the redistribution of welding residual stresses with fatigue crack extension occurs but also the phenomenon of the relaxation of welding residual stress with the fatigue loading might possibly occur at the same time and this would make the problem quite difficult [10]. Therefore, the effects of applied stress ratio (R_{app}) in correlation with residual stress ratio (R_{res}) and residual stresses distribution have to be investigated.

1.6. Objectives and Scope

From the design point of view, fatigue properties of welded structures such as the initial crack length, the final crack length, crack growth data and fatigue strength curve (*S-N* or Wöhler curve) should be determined accurately. Then, the fatigue life of welded structures can be correctly evaluated. Unfortunately, the real value of crack lengths and crack growth properties according to notch cases are not known sufficiently for most cases.

The welding standards contain descriptions of a number of possible weld geometries, with limits for the accepted dimensions of the defects. The fatigue life calculation of different weld classes as based on fracture mechanics approach requires fixed values of crack length and SIFs. The latter can be calculated due to applied load and due to residual stress.

Limited work is published for the calculation of SIFs under residual stress fields [11] and the effect of residual stresses on the crack growth life.

The aim of this work is to calculate the fatigue strength of notch cases using fracture mechanics based method and comparison with the solutions of the International Institute of Welding (IIW) [3], Eurocode 3 [12], Germanischer Lloyd Aktiengesellschaft (GL) [13] and British Standards Institution (BSI) [14].

The fatigue strengths have been calculated based on real values of crack length for each notch case that is calculated in this work. Moreover, the distributions of residual stresses and their influences on crack growth life were determined. In addition, the correlation between the residual stress distribution effect and load ratio R was investigated.

With the current fracture mechanics models, different notch cases could be defined including fatigue strength that is called FAT (fatigue strength at 2 million cycles) and residual stress distributions. The fatigue strength for unknown notch cases not listed yet in the standards can be calculated under the effect of different weld geometries and different residual stress profiles.

1.7. Outline

This study includes fatigue life calculations and comparisons with literature using fracture mechanics methods and is divided into six chapters.

Chapter One gives a brief *Introduction* to the main topics of the book.

In Chapter Two, a comprehensive *Literature Review* for fatigue life of welded joints, fracture mechanics approach and residual stresses is presented. The focus lies on LEFM (linear elastic fracture mechanics), which is used in this book. *The Modeling of Welded Joints and SIF Calculation* are thoroughly described in Chapter Three.

In Chapter Four, *Fatigue Life Calculations and Verifications*, analytical techniques and procedures that are used to predict the fatigue life of various structures are described in detail.

Data published in the open literature and the fatigue test results are presented as a part of comparisons and experimental verifications in the fourth chapter. Re-analysis and re-calculation of notch cases in the literature and standards are shown also in the fourth chapter. A discussion of *Residual Stresses in Welded Joints* is given in Chapter Five. Finally, *Conclusions and Recommendations for Future Work* are presented in Chapter Six. The evaluation of design codes is restricted to IIW, BS, Eurocode 3, and GL recommendations.

1.8. The Present Work

The fatigue testing of welded joints and large scale structures is time consuming and expensive. Therefore, the analytical procedures and software have been developed to predict the crack initiation, crack path and propagation time. However, the initiation time for cracks in weld joints is neglected in fracture mechanics approaches by supposing an initial crack length.

Linear elastic fracture mechanics can then be used to calculate the propagation portion of the total life by integrating the crack growth rate (da/dN)-stress intensity factor (ΔK) relation from a specific crack size to the critical crack size at fracture. However, the application of ordinary fracture mechanics parameters like K-factor to very small cracks (below 1 mm) is criticized (particularly in fatigue).

For each case the geometry is modeled and SIFs are obtained to calculate the fatigue life. The weld joint geometry with an initial crack is modeled in the FEM program FRANC2D. In this program a crack is propagated automatically or step by step according to the maximum stress direction. For every crack length, the SIFs were calculated. The results have been used to determine the SIF as function of crack length for the particular case in form of a polynomial function with suitable order. These FE results were benchmarked for the effects of mesh size, mesh type, mesh density, crack increment and effect of symmetry to evaluate the reliability of current FE mode in SIFs calculations.

The introduced function has been used for calculating the fatigue life by using fracture mechanics method. The life was assumed to be finished when the crack reaches half the sheet thickness; however, the final crack length has negligible effect on fatigue life. All steps of the procedure were described thoroughly in the following chapters.

The following points summarize procedures of the current approach:

1. FE modeling that gives accurate stress intensity factors.

2. With knowledge of the stress intensity factors at different crack depths, it was possible to make curve fits between SIF and crack length, then calculate $K_I(a)$ for different loading due to the linear relation between K_I and load. Also the effects of crack increment and mesh elements have been studied to give satisfactory results.

3. Insert the polynomial equation into fatigue life formula (da/dN-ΔK) and carry out the numerical integration to calculate the expected fatigue life of the specimens and construct the fatigue life curve (S-N).

4. Carry out the backward calculations for the fatigue life curve as compared with known cases from the standards for the value of fatigue strength life (FAT). Crack growth parameters are fixed which give coalescence between the fatigue life curve from the current calculations and those from standards.

5. Determine the initial crack length for different notch cases and reveal the typical length which would give satisfactory results for specific types of cracking joints and loading conditions. Crack growth parameters and final crack lengths have been fixed while the initial crack was manipulated. Final crack length was chosen to be equal to one-half of the plate thickness.

6. Investigate the effect of the residual stresses on the crack propagation life based on the calculated parameters from point 5.

7. Investigate the effect of applied stress ratio (R_{app}) and residual stress ratio (R_{res}) on fatigue life by incorporating them in addition to residual stress distributions into the fatigue life formula.

The incorporation all above steps in one approach was the task of the current work. These approaches are validated and compared through the following chapters of this work.

Chapter Two
OVERVIEW OF FAILURE OF WELDED JOINTS

2.1. Introduction

With welded joints, stress concentrations occur at the weld toe and at the weld root, which make these regions the points from which fatigue cracks may initiate. To calculate the fatigue life of welded structures and to analyze the progress of these cracks, fracture mechanics technique is used.

In most cases, the welded components are designed using S-N nominal curves which predict only the fatigue life of the component. However, in weldment, the fatigue crack propagation was more significant and takes a majority of total fatigue life.

Since the crack initiation occupies only a small portion of the life it can be assumed negligible. The fracture mechanics method is suitable for assessment of fatigue life and inspection intervals in weld structures.

2.2. Fracture Mechanics of Fillet Welded Joints

The inevitable parameter in fracture mechanics is the stress intensity factor (SIF) which is used in fatigue life calculation. SIF tightly knit with the fracture mechanics to predict the stress state near the tip of a crack caused by a remote load which should take into account the residual stresses in conjunction with geometry.

Nykänen et al. [15] investigated the fatigue behavior of 12 common types of welded joints parametrically and the tools which would allow more precise assessment of the effect of dimensional variations on the fatigue strength (FAT) were given [15]. They mentioned that in parallel joints toe cracks and lack of penetration (LOP) are frequently encountered defects in fillet welded joints. Toe cracks occur because of the stress concentration in the weld toe region, while LOP results from inaccessibility of the root region during welding.

In this Chapter, the summery of the 12 types of welded joints and the predicted FAT with the parameters are shown in Table 2.1. In these welded joints, the initial crack length equal to 0.2 mm was assumed using Paris' law numerical integration which is carried out automatically by the FRANC2D/L program [15]. This length is typical when arc welding is used. The root crack was varied depending on the degree of penetration. The different weld gaps between the two plates were assumed. Table 2.2 shows the Paris' constants and conditions for the 12 welded joints presented in Table 2.1, where C_{char} and C_{mean} refer to crack growth rate coefficient corresponding to 95% and 50% survival probability, respectively. No effect of residual stresses and their distributions were investigated in Nykänen's study.

Table 2.1: Common types of welded joints used in Ref. [15]

Joint types	Geometry	Predicted FAT (MPa)
Single fillet welded T-joint with partial penetration	50, 100, w, h, β, t, 100 mm, T	$FAT_{DOB} = \sum_{i=1}^{20} A_{DOB,i} \left(\frac{h}{t}\right)^{a_i} \left(\frac{w}{t}\right)^{b_i} \left(\frac{T}{t}\right)^{c_i} \cdot f(t)$

Joint type	Geometry	Formula
Corner joint with partial penetration		$FAT_{DOB} = \sum_{i=1}^{56} A_{DOB,i} \cdot \left(\dfrac{w}{p}\right)^{a_i} \cdot \left(\dfrac{h}{t-p}\right)^{b_i} \cdot \left(\dfrac{p}{t}\right)^{c_i} \cdot f(t)$
Angle joint		$FAT_{DOB} = \sum_{i=1}^{6} A_{DOB,i} \cdot \beta^{a_i} \cdot \left(\dfrac{t}{T}\right)^{b_i} \cdot f(t)$
Cruciform joint with V-butt welds and partial penetration		$FAT_{DOB} = \sum_{i=1}^{20} A_{DOB,i} \cdot \left(\dfrac{h}{t}\right)^{a_i} \cdot \left(\dfrac{w}{t}\right)^{b_i} \cdot \beta^{c_i} \cdot f(t)$
Cruciform joint with K-butt welds and partial penetration		$FAT_{DOB} = \sum_{i=1}^{20} A_{DOB,i} \cdot \left(\dfrac{h}{t}\right)^{a_i} \cdot \left(\dfrac{w}{t}\right)^{b_i} \cdot \beta^{c_i} \cdot f(t)$

Joint type	Diagram	Formula
T-joint with fillet welds		$FAT_{DOB} = \sum_{i=1}^{10} A_{DOB,i} \cdot \left(\dfrac{t}{T}\right)^{a_i} \cdot \left(\dfrac{h}{t}\right)^{b_i} \cdot \beta^{c_i} \cdot f(t)$
Transverse partial penetration butt weld		$FAT_{DOB} = \sum_{i=1}^{35} A_{DOB,i} \cdot \left(\dfrac{w}{t}\right)^{a_i} \left(\dfrac{h}{b}\right)^{b_i} \left(\dfrac{l_1}{l_2}\right)^{c_i} \left(\dfrac{b}{2l}\right)^{d_i} \cdot f(t)$
Transverse partial penetration single butt weld		$FAT_{DOB} = \sum_{i=1}^{35} A_{DOB,i} \cdot \left(\dfrac{w}{t}\right)^{a_i} \left(\dfrac{T}{t}\right)^{b_i} \left(\dfrac{h}{b}\right)^{c_i} \cdot (\tan\alpha)^{d_i} \cdot f(t)$
Non-load-carrying cover plate fillet welded joint		$FAT_{DOB} = \sum_{i=1}^{10} A_{DOB,i} \cdot \left(\dfrac{w}{l}\right)^{a_i} \left(\dfrac{h}{l}\right)^{b_i} \cdot \beta^{c_i} \cdot f(t)$
Butt weld with permanent backing bars		$FAT_{DOB} = \sum_{i=1}^{15} A_{DOB,i} \cdot \left(\dfrac{w}{t}\right)^{a_i} \cdot \left(\dfrac{t}{T}\right)^{b_i} \cdot f(t)$
Axially loaded lap joint		$FAT_{DOB} = \sum_{i=1}^{15} A_{DOB,i} \cdot \left(\dfrac{w}{t}\right)^{a_i} \cdot \left(\dfrac{t}{T}\right)^{b_i} \cdot f(t)$

Axially loaded symmetric double lap joint	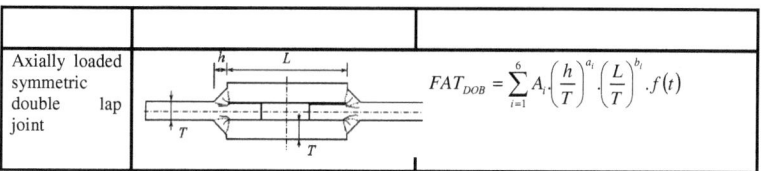	$FAT_{DOB} = \sum_{i=1}^{6} A_i \left(\frac{h}{T}\right)^{a_i} \cdot \left(\frac{L}{T}\right)^{b_i} \cdot f(t)$

Table 2.2: Material data for 12 welded joint reported in Ref. [15]

Initial crack	Material data	Approach
0.2 mm (toe crack) LOP (root crack)	C_{char} =3E-13, m=3 C_{mean}=1.7E-13, m=3	Paris' law numerical integration.

From literature [16-18], it is evident that most of the investigations on fatigue life prediction of the fillet welded joints are based on toe failure. Some other studies have considered the fatigue behavior of fillet welded joints failing from the root region.

Motarjemi et al. [19], Balasubramanian and Guha [20], Frank and Fisher [21], Usami and Kusumoto [22], have also studied the fatigue behavior and SIF of cruciform and T welded joints of carbon steels failing from the root (LOP).

Motarjemi et al. [19] evaluated SIFs at the crack roots of T and cruciform welded joints with LOP defects by FEM using ABAQUS package to analyze the different joint geometries. They found that the SIFs for cruciform welded joints were nearly always higher than those for the comparable T-welded joints.

Fatigue failure typically takes place at sites of high stress in either the base material or weldments. Weld toe contains the stress concentration site and small crack-like discontinuities [4, 23-26]. Such cracks tend to be along the line normal to the transverse stress.

Considering fatigue crack growth in welded joints, the percentage of the crack propagation phase in the total fatigue life depends very much on the quality of the weld comprising; weld geometry, initial defects in the weld, weld residual stresses and local stress conditions. Since welding defects can frequently exist in the vicinity of weldments, local stress concentrations around discontinuities and weld defects are fairly common. These crack-like defects begin to grow almost immediately when subjected to external cyclic fatigue loads, so that, for welded joints, the total fatigue life is mainly dominated by the crack propagation phase. Moreover, weld defects that were the source of crack initiation and growth were characterized. Internal discontinuities include porosity, entrapped oxides, and lack of fusion sites located in the longitudinal fillet welds and the groove welds [27].

Therefore, fracture mechanics approach assumes the existence of an initial crack a_i. It can be used to predict fatigue life and strength of the growth of the crack to its final size a_f. For welds in structural metals, crack initiation occupies only a small fraction of the life and it can be assumed negligible [8, 28]. Therefore, this method is suitable for assessment of fatigue life, inspection intervals, and crack-like weld imperfections that are likely present in weld joints. Initial cracks used in fatigue analyses are often in the range of 0.05-0.2 mm [29]. However, Engesvik [30] has also analyzed the fatigue life of welded joints and concluded that it may be dubious to apply LEFM at crack depths less than 0.1 mm. Nevertheless, this value can vary depending on the welding operation parameters, geometry and materials properties. Lindqvist [31] showed a 12 mm specimen's fracture surface after it was fatigue tested and afterwards broken up. Probably there were several small cracks along the weld toe. When the cracks grew in the direction normal to the applied load, they united into one large semi-elliptical crack. For load-carrying cruciform welded joints, lack of penetration (LOP) is considered to act as initial crack. Initial crack, a_i is usually measured or approximated to 0.1-0.2 mm for welds. Table 2.3 shows the parameters used in Lindqvist's study [31]

Table 2.3: Material data and geometry reported in Ref. [31]

Geometry	Initial crack	Material data	Approach
Cruciform weld joint with toe crack	0.15 mm (toe crack)	C_{char} =5.21E-13, m=3 C_{mean}=2.3E-12, m=3	Paris' law numerical integration.

BS7910 [14] recommended the initial flaw size a_i between 0.1 and 0.25 mm. The life is assumed to be finished when the final crack reaches half the sheet thickness [31, 32].

Nykänen et al. [15] mentioned that for toe cracks initially perpendicular to the plates an initial crack length a_i of 0.2 mm was assumed. They mentioned this length to be typical for arc welding and in case of LOP crack, the root crack was varied depending on the degree of penetration.

In fillet weld joints there are two locations where the crack may initiate and propagate (see Table 2.1). First at the weld toe and second from the root or LOP region. The root crack length or partial penetration depth (LOP) plays an important role in fatigue life. The first crack will propagate through the base plate whereas the second one will propagate through the weld throat [3, 33, 34].

Branco et al. [4] showed experimentally that the fatigue cracking for transverse butt, cruciform welds and non-load carrying fillet welded attachments usually initiated at a weld toe and propagated through the plate thickness. Maddox [35] stated that a fillet weld has small sharp defects along the weld toe from which fatigue cracks propagate. This effect combines with the stress concentration so that the fatigue life is effective in propagating the crack.

In most cases toe cracks have been considered [4, 31, 32, 36, 37] because they are easier to observe with the naked eye as well as with dye penetration tests. In addition to high stress concentration, the tensile residual stresses are located in this point. The tensile residual stresses of yield strength magnitude exist at the weld toe regions reducing the fatigue life [24].

Singh et al. [38] presented the findings of a study of the axial fatigue performance of AISI 304L load-carrying cruciform joints which failed in the weld metal with and without cryogenic treatment. The fatigue properties of cryogenically treated samples have shown improvement due to strain induced martensite that formed during cryogenic treatment and the associated generation of compressive stresses in the weld metal.

Hou et al. [39] mentioned that a crack depth of 0.25 mm was commonly used as terminate of crack initiation phase. Many strain gages placed along the weld direction in

T-joint, near the weld toe and the size and location of initiated surface cracks by was detected change in the strain gages readings. They claimed that cracks with a depth of 0.5 mm or even smaller could be detected. They recorded the total fatigue life of the T-joints and the crack propagation life from a crack depth of 0.5 mm to the final failure. They showed that fillet weld joints like T-joint have high stress concentrations at the weld toe. Therefore, it is easier to initiate a crack from the weld toe, hence, the value of initial number to total number of cycle (N_I/N_T) become smaller.

Balasubramanian et al. [20] analyzed the influences of two welding processes, namely, shielded metal arc welding (SMAW) and flux cored arc welding (FCAW), on fatigue life of cruciform joints containing LOP defects.

Finally, it should be mentioned that the lists of fatigue strength of fillet welded joints failing from toe and root cracks, respectively, are presented in the International Institute of Welding (IIW) [3, 34], Germanischer Lloyd Aktiengesellschaft (GL) [13], and British Standard Institution (BSI) [14].

2.3. Crack Propagation Curve

The fatigue of welded joints is a vast area and gives many possibilities for research. It has been found from experience that most common failures of engineering structures such as welded components are associated with fatigue crack growth caused by cyclic loading. Engineering analysis of fatigue crack growth is frequently required for structural design, such as in Damage Tolerance Design (DTD) and residual life prediction when an unexpected fatigue crack is found in a component of engineering structure. For analysis, the fatigue life of welded structures can be divided into two parts: crack initiation phase and propagation phase. The initiation life is defined by the number of loading or straining cycles, N_i, required to develop a crack of some specific size, a_i. The propagation stage then corresponds to that portion of the total cyclic life, N_p, which involves growth of that crack to some critical dimension at fracture, a_f. Hence $N_T=N_i+N_p$, where N_T, is the total fatigue life. Therefore, fatigue crack propagation behavior is typically described in terms of crack growth rate or crack length extension per cycle of loading (da/dN) plotted against the SIF range (ΔK) or the change in SIF from the maximum to the minimum load (see Figure 2.1) [28].

The first region (I) in Figure 2.1 is referred to as the near threshold region. It indicates a threshold value, below which there is no observable crack growth. The second region (II) is a linear region, which is known today as the Paris' law region. During this period, fatigue crack growth (FCG) corresponds to stable macroscopic crack growth. In the third region (III), FCG is very high as it approaches instability and point of fracture. There is little FCG life involved and it is primarily controlled by the fracture toughness (K_{IC}). The main part of the log-log plot of da/dN versus ΔK that is of most concern in most works is the Paris' region (Region-II). This region describes the crack growth behaviour using the relationship between cyclic crack growth rate da/dN, and the stress intensity range, ΔK [9, 40].

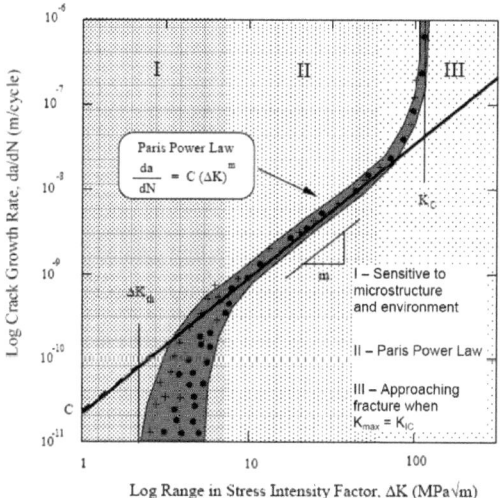

Figure 2.1: *The crack propagation rate versus the SIF range [40]*

The central portion of the crack growth curve (Region-II) is linear in the log-log scale. Linear Elastic Fracture Mechanics (LEFM) condition essentially deals with crack propagation in this region, which is commonly described by the crack growth equation proposed by P. C. Paris and F. Erdogan and popularly known as the Paris' law, was given in Ref. [28] as shown below:

$$\frac{da}{dN} = C(\Delta K)^m \tag{2.1}$$

where C and m are material dependent constants and ΔK is the range of SIF.

It should be noted that Paris' equation developed in 1963, only represents the linear phase (Region-II) of the crack growth curve [8]. As the SIF range increases approaching its critical value of fracture toughness (K_c), the fatigue crack growth becomes much faster than that predicted by Paris' law. Forman proposed the following relationship for describing region-II and III together [8]:

$$\frac{da}{dN} = \frac{C(\Delta K)^m}{(1-R)K_c - \Delta K} \tag{2.2}$$

where R is the stress ratio, equal to $\sigma_{min}/\sigma_{max}$.

Note that the above relationship (2.2) accounts for stress ratio, R effects, while Paris law assumes that da/dN depends only on ΔK. Based on the above relationship, fatigue crack propagation life can be predicted by integrating both sides of these functions if a suitable SIF solution is obtained.

It is to be emphasized that fatigue crack growth equation which is sensitive to R like NASGRO equation is recommended to use, or add to the Paris' model the crack closure phenomenon or any other model accounting for the state of affairs at the crack tip.

The equation is used in the most recent release of the crack growth prediction program, NASGRO. The NASGRO equation is written as:

$$\frac{da}{dN} = C_{ESA}\left[\left(\frac{1-f}{1-R}\right)\Delta K\right]^n \frac{\left(1-\frac{\Delta K_{th}}{\Delta K}\right)^p}{\left(1-\frac{K_{max}}{K_c}\right)^q} \qquad (2.3)$$

where C_{ESA} and n are empirical parameters describing the linear region of the curve (similar to the Paris' model), and p and q are empirical constants describing the curvature in fatigue crack growth rate (FCG) data that occur near threshold (Region-I) and near instability (Region-III), respectively. The Newman's effective stress ratio (f), the threshold value of SIF range for a given R, (ΔK_{th}) and the critical SIF (K_c) are presented in Chapter Four. The unit for the fatigue crack growth rate (FCG) da/dN is mm/cycle, and the SIF range ΔK is MPa (m)$^{1/2}$.

Research to date used the modified Paris' or Forman' law and incorporating them with residual stress intensity factor and total SIF to calculate lives which consider only the linear region of crack growth curve.

2.4. Simulation of Fatigue Crack Growth

In spite of the fact that several SIF handbooks have been published, it is still difficult to find solutions adequate to many welded configurations [27, 41]. This is mainly due to a wide variety of complex welded geometries, loading systems and the suitable solutions are not always available.

When no analytical solutions are available, several modeling methods may be used. The most modern ways to solve the SIFs are Finite Element (FE) or Boundary Element (BE) software. Automatic meshing during the crack growth is included in some software.

The finite element method (FEM) has been widely employed for solving linear elastic and elastic-plastic fracture problems. The evaluation of SIFs in 2D geometries by FEM is a technique widely used for non-standard crack configurations. As regards to through thickness weld toe cracks, no 3-D analysis results have yet been reported to the author's knowledge.

Over the years, the complexity of problems increased significantly and that makes it important to convert and provide a more useful method that can be used for fatigue problems.

Although the complexity of models has increased, the first applications were for relatively simple purposes such as the determination of SIFs for different crack configurations and different joint geometries.

Andersen [7] presented models of 2-dimensional crack simulation. In addition, the effects of residual stresses have been studied using the FE-based optimization program ODESSY. The stress intensity state at each location of the crack tip was used to predict the direction of the next crack increment. The procedure involved re-meshing of the domain containing the continuously changing crack geometry and a model for prediction of the crack growth direction. Thus, the procedure was used to simulate the crack growth. The obtained results showed a reasonable agreement with experimentally obtained data. The weakness of the approach was the missing ability of the re-meshing algorithm to produce a high quality mesh in the domain around the crack tip.

The main problem of using FE-based software is that the simulation of the crack growth is very time consuming. While the main problem using BE method is that the engineers are not familiar with this method and the solving of the BE model is not as fast as the FEM [42].

Caccese et al. [23] performed FE analyses on the cruciform-shaped test articles to ascertain the stress concentration factor for various shaped fillet geometries. The ANSYS^{TM} FE program was used for this purpose. Results of the FE analysis were processed using several analytical methods and these data were used to estimate the relative change in fatigue life due to the geometry of the fillet. Since it is desirable to use these methods with a multitude of weld profiles, the FE method was chosen for the analysis procedure to determine the stress concentration factor.

In ANSYS, to model the crack a small cut in the geometry is performed. This means that there is a relatively small distance between the points at the opening of the crack. In all models, the growth of the crack is perpendicular to the horizontal sheet. Crack growth is simulated by moving the crack tip key point into the material. Moreover, the crack can be difficult to mesh because the elements must be relatively small near the crack tip. In order to get around that problem the total area of the sheet is divided into smaller areas. Since the elements around the crack tip must be relatively small and larger elements are used far away from the crack it is extremely time consuming to use these small elements' model.

Problems may occur in the transition area between smaller and larger elements. This is done by creating a box around the crack tip, containing small elements. Although large efforts have been made to control the meshes, sometimes some bad elements appear around the box.

Therefore, Karlsson [32] found that using a smaller element side length on the global areas, 0.2 mm instead of 1 mm, the badly shaped elements disappear. The calculation time, however, is about 10 to 20 times longer and the calculated SIFs improve less than 1.5%. Since several geometries are to be calculated, the coarser mesh is used to save time.

Boundary Element Analysis System software (BEASY) [43] which is now a simple extension of the standard stress analysis is performed by engineers. Simply select the shape, size, and location of the crack and BEASY will automatically re-mesh the model to include the crack and compute the stress intensity data.

Moreover, BEASY not only provides stress intensity data but can also predict how the crack will grow and provide the automatic crack growth re-meshing which is automatically performed by BEASY where necessary. The procedure is highly automated with a crack growth wizard guiding the user. High accuracy historically, SIF and crack growth data have been approximated using standard reference solutions.

Fricke et al. [44, 45] performed BEM analysis using the code BEASY where the displacement method is used for the SIFs solution. The crack propagation has been calculated based on Paris' crack growth law using the computed SIFs along the crack front. It can be seen that the highest crack propagation occurs at two points namely at the end of the root gap and close to the corner.

2.5. Fracture Analysis Code

The FE program (FRANC2D) was developed by the Cornell Fracture Group from Cornell University, USA [46]. FRANC provides higher time saving and is more familiar with application and mesh generation.

The analysis was undertaken based on the assumption of an initially isotropic elastic material, in which a crack was subsequently allowed to form and grow according to a fracture criterion. Therefore, it combines with BEASY by the easy crack modeling and meshing.

In FRANC there is no reality for using the box which is used in ANSYS and there is not necessarily to use the fine mesh near the crack tip thereby saving time. It can be shown that FRANC has advantages over other FE software to calculate SIFs. It can be applied for mechanical component crack growing in addition to civil engineering applications. Moreover, FRANC2D can be used to represent the crack geometry in a fuselage, and to simulate crack interaction with various structural elements. However, Mashiri et al. [25] discussed the nature of 2-dimensional models and they assume that undercut is continuous throughout the whole length of the weld toe. This should give a conservative value of fatigue crack propagation life as compared to real cruciform joints under tensile cyclic loading where cracks can initiate at multiple points along the weld toe of the cruciform joints and propagate before coalescence occurs.

The simulation of fatigue assessment method by FEM using FRANC2D depends on SIFs. The analysis can be undertaken using a version of the program downloaded from the website and by using the support documentation provided.

Nykänen et al. [15] evaluated the two-dimensional finite element models of the joint using plane strain linear elastic fracture mechanics (LEFM) calculations. The as-welded condition was assumed with the result that no crack initiation period was considered and stress ranges were fully effective. In a wide variety of cases crack growth problems can be solved within the frame of linear elastic fracture mechanics (LEFM). This is the case when the yield zone at the crack tip is small with respect to both the crack size and the remaining ligament. By characterizing subcritical crack growth using the concept such as stress intensity factor K, it is possible to predict the crack growth rate of a weld under cyclic loading, and hence the number of cycles necessary for a crack to extend from some initial size, i.e., the size of pre-existing crack or crack-like defects, to a maximum permissible size to avoid catastrophic failures [15].

Hee et al. [47] showed that the FE program, FRANC2D is able to predict the growth of cracks for dam structure. The initially isotropic elastic material was assumed, in which a crack was subsequently allowed to form and grow according to a fracture criterion. In welded joints, the same assumptions are exploited.

FRANC2D program is highly interactive and has a user-friendly working environment. The Cornell Fracture Group [46] has also developed a more advanced FE program, FRANC3D for simulating 3-dimensional fracture growth. This program is capable of modeling multiple, non-planar and arbitrary shaped cracks as well as unstable cracks which have a non-linearity motion during the growth.

2.6. Factors Affecting the Fatigue Strength of Welded Joints

In welded structures, local stress concentration is inevitable due to structural geometry or discontinuity. Therefore, the fracture occurs at these stress concentration points.

Although fatigue crack growth are negative characteristics of materials, many factors have to be taken into account when fatigue crack growth data (curves) are to be measured or to be applied for fatigue life prediction, especially in welded joints [8]. Some of such factors are described below.

2.6.1. Weld Geometry

The local stress concentrations due to weld geometries and irregularities at the weld toe and crack-like defects are known to have influence on fatigue strength.

Fatigue life of a weldment is influenced by the material, environment, welding techniques, weld quality, connection details and the geometric profile of the weld. Welded joints are regions of stress concentration where fatigue cracks are likely to initiate [23]. Geometry is one of the primary factors that control the fatigue life. Accordingly, procedures that improve weld geometry profiles by reducing stress concentrations will have a beneficial impact on fatigue life. Most fatigue-life improvement methods implemented to date are post-weld operations.

Ferrica and Branco [48] investigated the effects of weld geometry factors on the fatigue properties of cruciform and T-weld joints. The results showed that the thickness of main plate and the radius of weld-toe are the most important factors for the fatigue properties of welded joints.

Kainuma and Mori [49] carried out experimental fatigue tests on load-carrying welded cruciform joint specimens to study the effect of weld shape on fatigue strength of load-carrying cruciform joints failed from root. They showed that the fatigue crack originated from the weld root and propagated into the weld metal in a direction roughly perpendicular to the applied load. This direction coincided with that of the unwelded line. A failure line leaning approximately 45° toward the unwelded line was generated by the final and static failures. A fatigue crack propagation analysis for joints with various weld shapes, plate thickness values, weld sizes, and weld penetration depths was also performed, and a quantitative fatigue strength evaluation method for the joints was derived. Kainuma et al. [49] found that the plate thickness has a more remarkable effect on the fatigue strength in non-load carrying cruciform fillet welded joints and in joints where a fatigue crack initiates from the weld toe. In these cases, the fatigue strength decreases inversely to the 1/4th power with the plate thickness. As demonstrated above, the degree of the thickness effect in non-load carrying cruciform fillet welded joints differ from that in load carrying joints that fail from the weld root. This is because the thickness affects different mechanisms in the two joints. When a fatigue crack originates from a weld toe, the thickness affects the stress concentration at the weld toe, but when a fatigue crack originates from a weld root, the thickness affects the initial crack size.

Kainuma and Mori [49] also studied the effect of five different weld shapes, an isosceles triangle, scalene triangles with longer leg lengths on either the main plate side or the cross plate side, and concave or convex curvatures, in order to determine the influence of the weld shape on the fatigue strength. They stated that one type of fatigue cracks originates from the weld roots of load-carrying fillet welded cruciform joints. The main conclusions were the fatigue strengths of joints with scalene-triangle-shaped welds with a longer leg length on the cross plate side, joints with concave welds, or joints with convex welds were similar or slightly greater than those of joints with isosceles triangle-shaped welds for a given weld size. Compared with the fatigue strength of these joints, the fatigue strength of joints with a scalene-triangle-shaped weld with a longer leg length on the main plate side was high. As the weld penetration depth increased, the fatigue strength also increased. This latter effect was evaluated by applying the idea that the weld penetration depth reduced the plate thickness.

Nguyen et al. [50] studied the effect of important butt weld geometry parameters e.g. tip radius of undercut at weld toe, weld toe radius, flank angle, plate thickness and edge preparation angle, and the effect of initial crack geometry on the fatigue crack propagation life by using LEFM. A simple mathematical model has been developed to predict the co-influence effect of the above-mentioned weld geometry parameters on the fatigue stress range *vs* life (S-N) curve. This model gives an explanation for the overall effect of weld geometry parameters as the main reason of scatter phenomenon in fatigue testing practice.

Nguyen et al. [37] showed that the fatigue life of the welded joint is increased as the weld toe radius increases. Moreover, the fatigue life of the welded joint is increased by decreasing the values of flank angle. They have shown that significant effects occur in specified range for weld toe angle and toe radius. Outside these ranges, insignificant improvement is obtained. Weld toe radius and toe angle are kept constant to study the effect of plate thickness on SIF for transverse butt weld. It shows that the value of SIF is increased as the value of plate thickness increases. This can explain the well known effect of plate thickness in reducing the fatigue life of a weld joint. The fatigue life is significantly increased as the plate thickness decreases from 32 mm to 9 mm. However, the improvement of fatigue life due to the decrease of plate thickness from 20 to 9 mm is insignificant. It means that for the lower range of plate thicknesses (less than 20 mm), the effect of plate thickness is ignorable.

Caccese et al. [23] evaluated the effect of weld geometry profile on fatigue life of laser-welded HSLA-65 steel. Results are presented of cruciform-shaped fatigue specimens with varying weld profiles loaded cyclically in axial tension–compression. Better quality control of weld geometry is possible with laser welding compared with more conventional techniques such as SMAW and GMAW. They have shown that the use of a combined laser and gas-metal-arc welding (GMAW) weld procedure results in a substantially improved geometric profile of a longitudinal fillet weld. Welds with improved geometric profile can result in much better fatigue life than the same size weld with other profiles.

Branco et al. [4] showed that fracture mechanics modeling confirmed that the differences in weld toe geometry were consistent with the differences in fatigue life actually observed. Weld toe radius appeared to be more significant than weld toe angle.

It has been known that fatigue crack growth behaviors of welded joints highly depend not only on the materials and load conditions but also on weld geometry such as weld toe angle, θ (where $\theta=180-\gamma$), weld toe root radius (ρ), plate thickness (t), width of the weld bead (W), height of the weld bead (H), and (L) is the plate length as shown in Figure 2.2 [8].

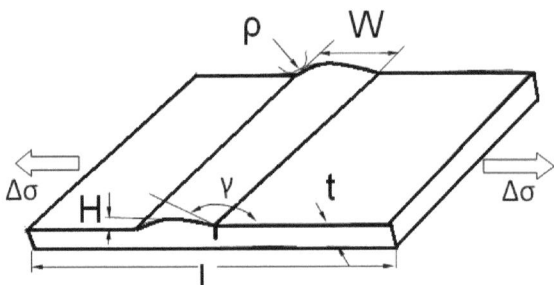

Figure 2.2: Weld geometry parameters at butt-weld joints [8]

Ferrica and Branco [48] investigated the effects of weld geometry factors on the fatigue properties of cruciform and T-weld joints. The results showed that the ratio of thickness of main plate and the radius of weld-toe (t/ρ) are the most important factors for the fatigue properties of welded joints. As the thickness of plates is reduced, the fatigue strength and life will increase in the case of T and cruciform joints. These results have been confirmed by other researchers [51, 52] studied GMAW as welded butt welds in 25-30 mm AA5083-O, AA5083-H116, AA5456-H116 and AA5456-H117 aluminum alloys. The investigations concluded that the angle at the toe of the reinforcement is the most critical factor in the determination of the fatigue life of as welded joints and actual maximum stress at the weld toe.

The removal of weld reinforcement may increase the fatigue life, whether exposed to air or seawater, if care is taken to prevent surface defects during the bead-removal process. Niu and Glinka [53] derived an empirical relation, accounting for the stress concentration K_t at weld toe, by FE methods [8, 28] as follows:

$$K_t = 1 + 0.512 \theta^{0.572} \left[\frac{t}{\rho} \right]^{0.469} \qquad (2.4)$$

where θ is the weld toe corner (then $\theta = 180 - \gamma$) angle in radian, t is the plate thickness, and ρ is the weld toe root radius (see Figure 2.2).

Figure 2.3 shows the stress concentration factor as a function of weld angle and thickness for a weld toe radius equal 1 mm. It was shown that the stress near a notch tip is dependent mainly on the stress concentration factor K_t, the notch tip radius, ρ, and the weld toe corner angle θ. Therefore, it can be concluded that any changes in the weld toe radius mostly affect the stress concentration factor K_t, and the stress field in the close vicinity of the weld toe.

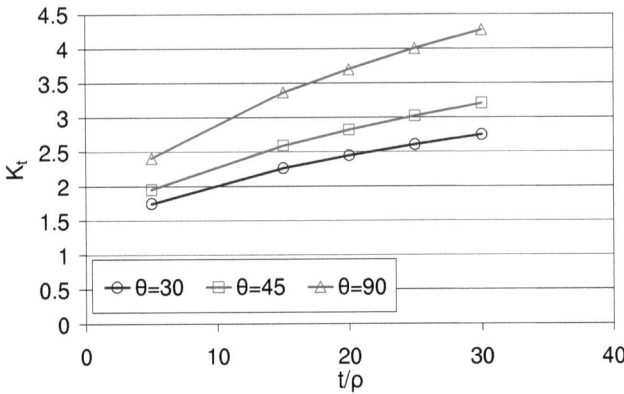

Figure 2.3: Stress concentration factor as a function of weld angle (radian) and thickness (mm) for weld toe radius equal to 1 mm

The welds with their reinforcement removed were more sensitive to porosity compared with the weld with reinforcement intact, although the former gave longer life than as welded joints when porosity level was lower. In other words, when a relatively higher level of weld inner-defects such as porosity, inclusions, lack of fusion and LOP, exist in a weldment, welds with bead reinforcement removed may exhibit shorter life than those of welds with bead on. This situation could be fairly true for weldments with a high level of weld inner-defects, for example weld repaired structures. Therefore, the as-welded joints can exhibit different fatigue crack growth performance behavior compared to that of the weld with bead reinforcement intact, depending on the situation.

Motarjemi et al. [54] studied for cruciform joints the effect of attachment thickness, main plate thickness, weld leg length, and initial LOP size on fatigue life by integration of the Paris' law. For T-welded joints, the effect of different support span has also been investigated. The results are reported as fatigue life curves for both types of joints. All the curves show higher fatigue life for T-joints than for cruciform joints. The studied parameters include the attachment and main plate thickness (B and T, respectively), weld leg length (h), support span (D) (for the T joints only) and initial lack of root penetration size ($2a_i$). The details of the parameters covered in this study are shown in Figure 2.4.

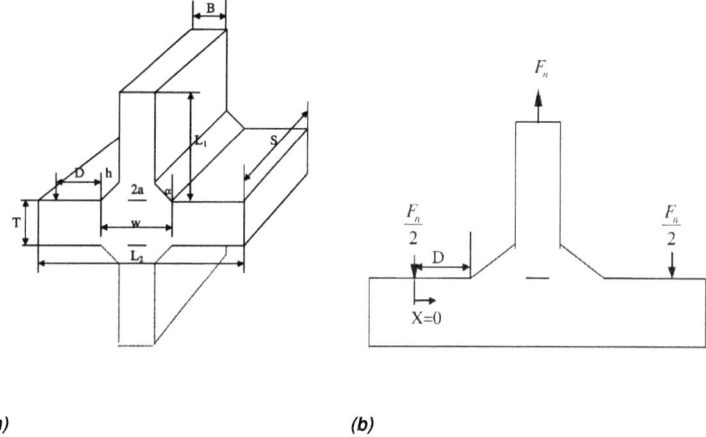

(a) (b)

Figure 2.4: *(a) Cruciform and T welded joint geometry; (b) simply supported condition* [54]

Taylor et al. [55] studied the thickness effect and showed that the weld in thicker section have short fatigue lives at the same stress nominal. Thinner plates have higher fatigue life than thicker plates.

Lindqvist [31] performed facture mechanics analyses using FEM and according to the requirements in the British Standard to calculate the fatigue life. The FEM analyses indicated that the fatigue life for the 6 mm specimen was about twice that of the 12 mm specimen.

The National Institute for Materials Science (NIMS) [56] presented a data sheet on fatigue properties of non-load-carrying cruciform welded joints of SM490B rolled steel for welded structure under different plate thickness (see Figure 2.5). These data showed that the fatigue life increases with decrease the plate thickness.

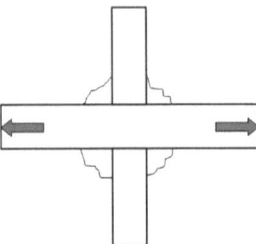

Figure 2.5: Non-load carrying cruciform welded joint

Branco et al. [4] used the TIG and plasma welding in the joining of thin sections, less than 10 mm thickness. This may introduce an additional benefit from the fatigue viewpoint, since fatigue strength is expected to increase with a decrease in plate thickness. Superior fatigue performance was confirmed for TIG and plasma transverse butt, cruciform and non-load carrying fillet welds of carbon-manganese steel, all joints failing from the weld toe. Weld details, which failed by fatigue cracking in the weld throat, showed no influence of welding process. Fracture mechanics modelling confirmed that the differences in weld toe geometry were consistent with the differences in fatigue life actually observed. The details of the test specimens are given in Figure 2.6. They showed that the radius appeared to be more significant than weld toe angle.

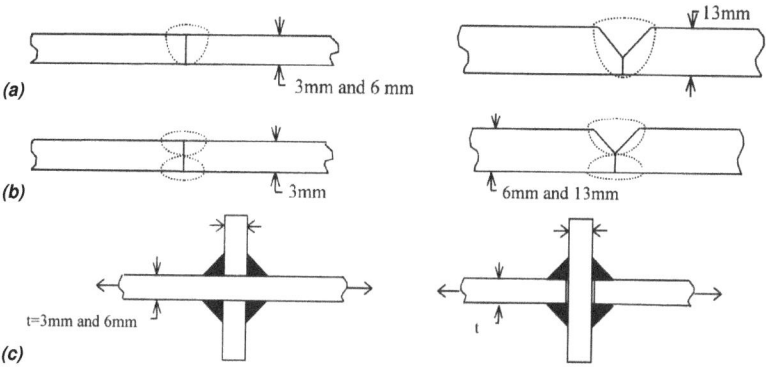

Figure 2.6: Details of the tested welded joints: (a) butt welds from one side (TIG and plasma); (b) butt welds from both sides (MMA and 6mm TIG); (c) cruciform fillet welded specimens designed to fail in a plate at weld toe (non-load carrying attachments) and failing a weld throat (load carrying) [4]

2.6.2. Weld Defects and Metallurgy

Weld defects are those imperfections or discontinuities produced in the weldments because of the weld process, such as porosity, LOP, slag inclusions, incomplete fusion, misalignment, undercut, weld profile etc. These weld defects can significantly influence the local stress field in the vicinity of welds when the welded component is subjected to cyclic fatigue load. In most cases, weld defects lead to severe stress concentrations and thus accelerate fatigue crack growth. Several works studied the effects of LOP on the fatigue behavior of welds. They concluded that LOP defects can seriously reduce the fatigue life of both types of weld, those with the reinforcement intact and those with the reinforcement removed. They reported that the effects of internal discontinuities on fatigue performance of welds with reinforcements are minimal. The effect of weld reinforcement is so marked that only critical defects would affect the fatigue behavior and this is related to the loss of cross section area. Porosity only becomes a factor when the reinforcement is removed.

Weld toe cracks are often found in many important welded structures. The ability to assess the effects of these defects on fatigue life of the welded structures is therefore of practical significance. To carry out this assessment, it is necessary to have reliable SIF solutions.

2.6.3. Materials and Welding Techniques

Chemical composition and condition of welding materials, including base metal and fillet material can directly affect fatigue crack growth rates of welded joints [57]. Different materials have different mechanical properties and characteristics under cyclic loading. Phase change will occur in most kinds of steels when subjected to the welding process and residual stresses are introduced by these phase changes in areas like the weld toe and heat-affected zone. Generally these residual stresses tend to be tensile stress on hot spots and hence reduce the fatigue performance of welded joints.

The magnitude of this influence is dependent on the type of material. However, in the case of austenite stainless steel and 5083 aluminum alloys, no phase change occurs and hence no residual stresses due to phase change exist, although the latter may suffer substantial strength loss due to heating during the weld process. Different welding processes also have different effects on the fatigue properties of welded joints. Some welding methods such as submerged arc welding cause heavy heat input in the weld joints, producing higher thermal residual stresses. As mentioned previously, defects introduced during the weld process such as undercut, lack of fusion, lack of penetration, misalignment, porosity and poor profiles can also have a significant influence on the fatigue properties of welded joints [8].

2.6.4. Weld Residual Stresses

Residual stresses are known since many years as one of the most critical factors which can significantly influence the fatigue properties of welded joints. Many researchers have reported on the basis of their experiments that the magnitude of the highest tensile residual stress may approach the yield strength of the parent material in some critical locations of weldments [5].

Previous investigations [24, 37] indicated that the WRS have also significant effects on the fatigue crack initiation phase and the early stage of crack propagation.

2.7. Residual Stress Effects on Fatigue Life of Welded Joints

Residual stresses are stresses that remain in components after the applied external load is removed. Residual stresses in engineering structures occur from welding and other manufacturing processes such as rolling, forging and casting. These stresses have a significant influence on the fatigue lives of structural components [5].

Weld residual stress (WRS) introduced by the welding process can come from the expansion and shrinkage of weldments during heating and cooling [24], misalignment and microstructure variation in weldments and heat-affected zone (HAZ). Residual stresses in weldments have two effects. Firstly, they produce distortion, and second, they can be the cause of premature failure especially in fatigue fracture under lower external cyclic loads [8].

It has been confirmed that WRS is quite large and will be tensile in the vicinity of the weld, where their magnitude is approximately equal to the yield strength of the weld metal, as shown in Figure 2.7.

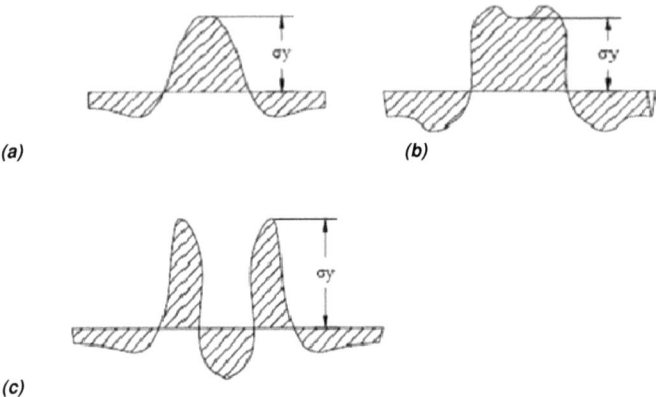

Figure 2.7: Typical longitudinal WRS distribution at butt weld [28]: (a) mild steel; (b) aluminum alloy; (c) high alloyed structural steel [8]

Some researchers confirmed that the tensile residual stresses could significantly decrease the fatigue properties of welded joints. It has been shown that tensile residual stresses in welded structures can be as high as the yield strength of the material and have a detrimental effect on the fatigue behaviour. On the other hand, compressive stresses on the surface of weldments could have a favorable effect on fatigue life and they can significantly improve the fatigue strength of welded structures [5, 9, 59].

Itoh et al. [58] studied the effect of residual stress on fatigue crack propagation rate in longitudinal welded residual stress field and found that the effect of crack growth rate in a weld residual stress field could be evaluated in terms of the effective stress intensity range, based on the measurement of effective stress ratio and crack opening ratio. The effective SIF and the effective stress ratio can be applied to predict the fatigue crack growth rate in both tensile and compressive residual stress field by using base material's crack growth rate data with different stress ratios.

Other works investigated residual stresses on the crack initiation phase only such as Teng et al. [24], who studied the effect of residual stresses on the fatigue crack initiation (FCI) life in high cycle fatigue of butt welded joints. They found that the localized heating caused by welding and subsequent rapid cooling can cause tensile residual stresses at the weld toe of butt-welded joints. These tensile residual stresses were considered one of the major influences on fatigue strength.

O'Dowd et al. [60] determined the SIF value for cracks of different sizes at the weld toe due to the weld residual stress using the FEM. They also compared the SIFs for the stress distributions provided in R6 and BS7910 and from neutron diffraction measurements along the weld toe line which determines the stress field directly from the measured elastic strain for three mutually perpendicular stress components (transverse, longitudinal and normal stress), see Figure 2.8 The three stresses are obtained from three strain components using Hook's law. Very dense mesh has been used near the weld toe of the T-plate weld joint. All calculations were carried out using the commercial FE software package, ABAQUS.

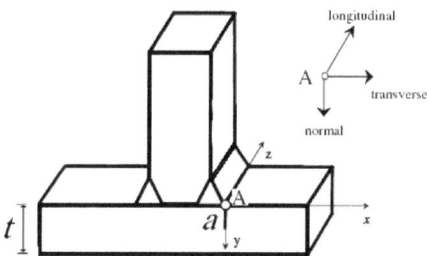

Figure 2.8: Geometry of T-plate and weld [60, 61]

2.7.1. Crack Closure and Effective SIF

When a specimen is cyclically loaded between K_{max} and K_{min}, the crack faces are in contact below K_{op}, the stress intensity factor at which the crack opens. Figure 2.9 illustrates the crack closure concept and shows that the portion of the cycle that is below K_{op} does not contribute to fatigue crack growth. The effective SIF range is defined as follows [28]:

$$\Delta K_{eff} = K_{max} - K_{op} \tag{2.5}$$

The effective stress intensity factor ratio is [28]:

$$R_{eff} = \frac{\Delta K_{eff}}{\Delta K} = \frac{(K_{max} - K_{op})}{(K_{max} - K_{min})} \tag{2.6}$$

The modified Paris-Erdogan equation is derived as [28]:

$$\frac{da}{dN} = C(\Delta K_{eff})^m \tag{2.7}$$

Eq. (2.7) has been reasonably successful in correlation of fatigue growth data for various metallic materials at different R ratios.

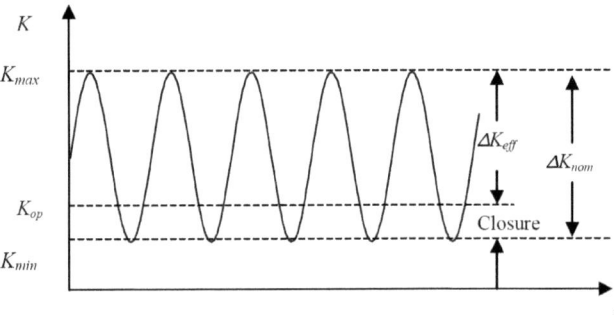

Figure 2.9: Definitions of effective SIF and crack closure concept [28]

In case of residual stress, two approaches, introduced by Glinka and Elber edited by Wu [28] have been frequently employed to account for the effect of residual stress on fatigue crack growth via crack closure. One method uses residual SIF ratio as follows [6, 28]:

$$R_{res} = (K_{min} + K_{res})/(K_{max} + K_{res}) \neq R_{app} \tag{2.8}$$

$$\Delta K_{eff} = K_{max}(1-R) \tag{2.9}$$

Here $K_{min,app}$, $K_{max,app}$ and K_{res} denote the values of the minimum and maximum SIFs due to the applied loading, and the SIF introduced by the residual stress field respectively. R_{res} is defined as the residual stress ratio. Nominal applied stress ratio R or R_{app} is the ratio of K_{min}/K_{max}. This approach requires the residual stress distribution and is associated with the SIF calculation which is also dependent on the fatigue crack length. It has been known that the welding residual stresses are distributed in a considerably non-uniform manner and are redistributed during the process of cyclic loading and fatigue crack growth. Therefore, Wu [28] mentioned that it is difficult to apply this approach for effective SIF evaluation because of the difficulty in capturing the actual welding residual stress (WRS) field during fatigue crack growth. The other method is to use the effective stress intensity factor range (ΔK_{eff}) determined from the crack opening ratio (R') as [28]:

$$\Delta K_{eff} = \Delta K_{nom} \times R' \tag{2.10}$$

where,

$$R' = (\sigma_{max} - \sigma_{open})/(\sigma_{max} - \sigma_{min}) \tag{2.11}$$

with σ_{open} being the measured crack opening stress in the residual stress field. In applying this approach, accurate determination of the σ_{open} is necessary for the effect of WRS and crack closure to be correctly estimated.

2.7.2. Determination of Residual Stresses

Residual stresses can be measured by different techniques classified into destructive, semi-destructive and non-destructive measuring methods (mechanical, optical, ultrasonic, X-ray or neutron diffraction and analytical).

Many researchers have pointed out that, for welded joints, weld geometry factors and weld residual stress (WRS) can significantly affect the final crack growth data in the corresponding fatigue experiments, see Ref. [9].

Assis et al. [62] presented the experimental results of residual stress measurements by X-ray tensiometry. Both longitudinal and transverse stress components (parallel and perpendicular to the weld seam, respectively) were measured in the seam weld, the heat-affected zone (HAZ) and in the base metal near the weld region, see Figure 2.10.

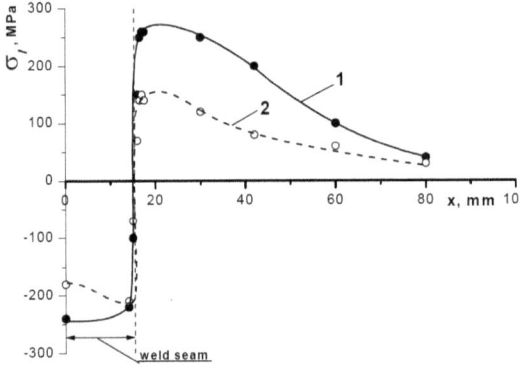

Figure 2.10: Residual stress distributions on the front face of the welded plate with linear weld seam: 1-longitudinal stresses; 2-transverse stresses [62]

The high tensile residual stresses are induced in weld metal as it contracts during cooling and have a detrimental effect on fatigue life under cyclic loading [59, 63].

The presence of compressive residual stress in the weld metal was able to increase the fatigue strength because compressive residual stresses have a beneficial effect on fatigue life [5, 38, 44, 45, 59, 64].

Shen and Clayton [63] studied the fatigue behavior of fillet welded ASTM A515 steel. They mentioned that under tension-tension cyclic load conditions, residual stresses did not show a significant effect on the fatigue strength of fillet welded specimens. Under tension-compression cyclic load conditions, the stress-relieved specimens showed a significant increase in fatigue strength over the as-welded specimens.

Cordiano [64] made a determination of the influence of various mechanical finishing procedures on residual stresses. He had measured the residual stress by hole drilling.

Under pulsed tension (tension-tension) when high stress concentration factor, there was little difference in crack propagation for un-welded, as welded and stress relieved specimens. In this case of tension-tension, no effect of residual stress on fatigue life was found, while in tension-compressive cyclic loads, the stress relieved specimens exhibited superior fatigue life. Cordiano [64] also found that the tensile residual stresses do not have a significant effect on fatigue life of all type of used pulsating load (tension stress or zero to maximum tension range). While under reversed cyclic load, the effect becomes apparent and increases with decrease the stress range to some limit.

Nguyen et al. [37] studied the effect of residual stresses and weld geometry on the fatigue life of butt welded joints by developing an analytical model using LEFM, superposition and FE approaches. They aimed to find the theoretical explanations for the improvement of fatigue life due to the compressive residual stresses introduced by various post-weld surface treatments and improved weld geometry because of the theoretical analysis of the effect of compressive residual stresses and weld geometry on the improvement of fatigue life of welded structures is still not clear. The residual stress distributions are not measured but assumed based on numbers of studies related to residual stresses in as-welded and shot peened welded joints.

O'Dowd et al. [60] obtained the residual stresses in welded ferritic steel T-plates, determined using neutron diffraction.

The measured data for the T-plate are shown in Figure 2.11. It has been found that the transverse residual stress distribution for different plate sizes and yield strength are of similar shape and magnitude when normalized appropriately and peak stresses are on the order of the material yield strength. The magnitude of the peak transverse stress is about 450 MPa (approximately 60 percent of the material yield strength) and occurs at a distance, y, about 5 mm from the weld toe (see Figure 2.11). SIFs corresponding to this residual stress field due to a crack of varying size at the weld toe will be calculated and discussed in Chapter 5 in relevant comparisons between the current work and the Refs. [60, 61].

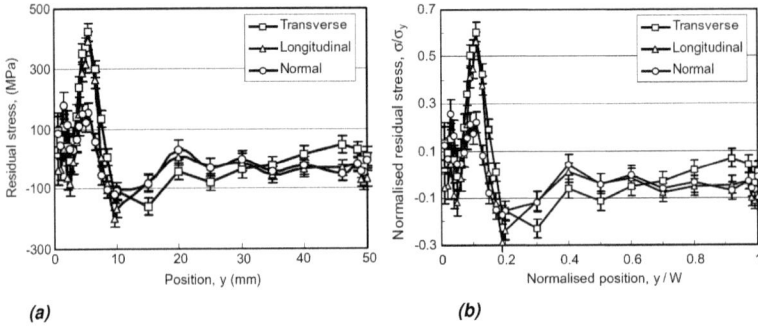

Figure 2.11: *Residual stress distributions for the high strength steel T-plate, Grade SE702 [10]: (a) stress in MPa and (b) normalised stress [60, 61]*

Servetti et al. [65] predicted fatigue crack growth rate in welded butt joints. They presented a simple method for predicting fatigue crack growth (FCG) using base material coupon test data and the effective stress ratio to account for the residual stress effect. FEM was used for calculation the SIF due to the applied and residual stress field.

2.8. Superposition Method

A quantitative assessment of the influence of a residual stress field on failure can be made by applying the principle of superposition to calculate an effective stress intensity factor, K_{eff}.

Stacey and Webster [66] quantified the influence of residual stress distributions on the SIFs developed in thick-walled cylinders containing line cracks and they evaluated the SIF caused by residual stress field, K_{res} using the superposition principle.

Nguyen, Wahab and Maksymowicz et al. [37, 67, 68] developed an analytical model using LEFM, FEA, dimensional analysis and superposition approaches to determine the effect of residual stresses, weld geometry and undercut on the fatigue life of butt welded joints.

Ramesh [6] found that the effects of the compressive residual stresses obtained from overloads and cold expansion of holes are included in the calculation of fatigue crack propagation by using the superposition method. The superposition method can account for residual stresses by a superposition of applied and residual SIFs, K_{app} and K_{res} respectively, to give the total SIFs, K_T under mode-I conditions [6] i.e.:

$$K_T = K_{app} + K_{res} \qquad (2.12)$$

With the assumption that the applied cyclic stress does not significantly alter the state of the residual stress during crack propagation, the total SIF range is [6]:

$$\Delta K_T = (K_{max})_T - (K_{min})_T = (K_{max,app} + K_{res}) - (K_{min,app} + K_{res})$$
$$= K_{max,app} - K_{min,app} = \Delta K_{app} \qquad (2.13)$$

and R_{res} is the residual SIF ratio as stated in Eq. (2.8).

The SIF range does not change since the SIF from the residual stress is negated, but the residual SIFs have an effect on the stress ratio. Hence, fatigue crack growth is predicted using the following correlation [6]:

$$\frac{da}{dN} = f(\Delta K, R_{res}) \qquad (2.14)$$

2.8.1. Residual Stresses Distribution

Profiles for residual stresses are presented in structural integrity assessment procedures for European industry (SINTAP). The aim of SINTAP document (Task 4) [69] is to propose a single reference source for residual stress profiles taken from open literature and SINTAP project. The profiles presented in SINTAP are more realistic than those proposed in the recent codes (except BS 7910 [14]) but still conservative [69].

Lee et al. [70] carried out reviews of through thickness transverse residual stress distribution measurements in a number of components, manufactured from a range of steels. They considered the residual stresses introduced by welding and mechanical deformation. The geometries consisted of welded T-plate joints, pipe butt joints, tube-on-plate joints, tubular Y-joints and tubular T-joints as well as cold bent tubes and repair welds.

Lee et al. [71] determined the residual stress distributions for plate T-butt welds from a detailed FEA of the welding process and they were compared with those of the measured data for validation. The residual stress distributions from the analyses and measurements were shown to be in similar shape. The distributions were found to be below the master curve (upper bound curve) for the residual stresses that were previously determined from a statistical analysis for a range of weld geometries and materials. A failure assessment for the T-butt weld with cracks under residual stress distributions has been carried out.

2.8.1.1. R6 Distributions for T-Plate

In the R6 procedure, two approaches for defining residual stress profiles in welded T-plates are provided, depending on the available information about welding conditions [61]. If the welding conditions are known or can be estimated, then the residual stress profiles will be given according to the size of the plastic zone (r_0) [61]. If the welding conditions are unknown, then BSI7910 polynomial distribution functions are provided (see Eq. 2.15). More details are found elsewhere [61, 69].

2.8.1.2. Polynomial Distributions for T-Plate and Butt Welds

If weld informations is unavailable then the BS7910 polynomial distribution is used [61]. BS7910 provides two transverse residual stress distributions for T-plate joints. The first transverse residual stress distribution is a polynomial function representing an upper bound fit to experimental data and is given by Eq. (2.15) as follows [69, 70]:

$$\sigma/\sigma_y = \left(0.97 + 2.3267(a/t) - 24.125(a/t)^2 + 42.485(a/t)^3 - 21.087(a/t)^4\right) \quad (2.15)$$

The transverse through thickness profile of butt welds is given as follows [69]:

$$\sigma/\sigma_y = \left(1 - 0.917(a/t) - 14.533(a/t)^2 + 83.115(a/t)^3 - 21.545(a/t)^4 + 24416(a/t)^5 - 9636(a/t)^6\right) \quad (2.16)$$

This distribution is here referred to as BS7910 [107]. The second distribution follows that in R6, with the distribution depending on the size of the plastic zone [61]. The distance r_o represents the size of the yielded zone (mm).

If $r_o \leq t$, where t is the plate thickness (mm) as shown in Figure 2.8, then:

$$r_o = \sqrt{\frac{k}{\sigma_y} \frac{\eta q}{v}} \quad (2.17)$$

where k is a material constant that depends on the coefficients of thermal expansion, Young's modulus, density and specific heat of a material (Nmm/J), σ_y is yield or 0.2% proof strength of the parent metal, q is arc power, v is weld travel speed and η is process efficiency (fraction of arc power entering plate as heat). Typical values of k and η for a range of materials are provided in the R6 documents. For ferritic steels the values provided are $k = 153$ Nmm/J, and $\eta = 0.8$, more details are found in Ref. [61].

When the plastic zone size r_o calculated by Eq. (2.17) is less than the base plate thickness, the residual stress is taken to be that of the parent material yield stress level at the weld toe, reducing linearly to zero over the size of the yielded zone as in R6. If Eq. (2.17) results in a plastic zone greater than the base plate width, the stress is taken to be equal to the yield strength across the whole specimen thickness [61].

This distribution as a function of the welding conditions and of the mechanical properties of the materials was preferred because it is less conservative than the first one (polynomial function from experimental measurements) [69].

The recommended transverse residual stress profile for the tubular T-joints is the same polynomial function as that provided for T-plate welds (Eq. 2.15). Lee [61] mentioned that the BS7910 provides a more conservative residual stress profile than R6 for the T-plate while R6 is more conservative for the tubular T-joint.

2.8.2. Residual Stress Intensity Factor

Fracture assessment for cracks will generally require the linear elastic SIF due to the weld residual stress and variation of weld geometry and any additional primary mechanical loading.

It is well known that fatigue performance of a welded structure depends upon local effects, such as local stress fields, defect conditions and material properties. Since the defects of welded structures are in most cases so severe, the crack initiation phase can be neglected.

LEFM principles can be used to evaluate the fatigue crack growth behavior and thus, to predict fatigue life of welded structures. In order to appropriately assess fatigue crack growth process in welded joints it is necessary to obtain accurate results for SIF solutions in the crack propagation phase [8]. Generally the SIF for a crack in a welded joint depends on the global geometry of the joint which includes the weld profile, crack geometry, residual stress conditions and the type of loading. Therefore, the calculation of the SIF, even for simple types of weldments, requires detailed analysis of the several geometric parameters and loading systems. The two approaches that have mostly been used till now for assessing SIFs for crack in weldments are weight function method and the FEM. The weight function method is an analytical technique for deriving SIFs from knowledge of the stress distribution in the un-cracked body. The numerical approach is based on basic weight functions applied for SIF calculation:

$$K = \int_0^a \sigma(x) m(x,a) dx \qquad (2.18)$$

Usually in this equation, $\sigma(x)$ is the stress distribution over the crack depth center line for a body without cracks. The weight function $m(x, a)$ is a unique property for a given body geometry.

The stress distribution, $\sigma(x)$ can be calculated by FE method. Weight functions $m(x, a)$ have been derived by Bueckner and Rice which are presented in Ref. [53] for 2-D and 3-D model edge crack and surface semi-elliptical crack in finite thickness plate, respectively.

Equation (2.18) reflects the fact that when a crack slits over a highly-stressed area (i.e., $\sigma(x)$ is high), the material ahead of the crack front will act as an alternative stress path, i.e. it will be highly-stressed locally. The results from a 3-D analysis are shown in Figure 2.12 [1]. F(a) is shown for a plane plate and T-butt joint.

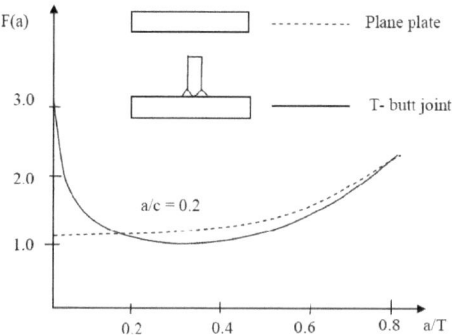

Figure 2.12: *Geometry functions for semi-elliptical surface cracks in plates and in T-butt joints [1]*

Concerned about the effects of weld profile geometry factors on local SIF evaluation, Niu and Glinka [53] developed weight functions for 2-D models with edge and surface semi-elliptical cracks in flat plates and plates with corners. Although the derived functions exhibited validity against available literature data it is still an approximation method because the 3-D nature of weld geometry is ignored. Based on the Bueckner's weight function, Ngugen et al. [72] created a semi-elliptical crack model for the SIF calculation on butt weld joints considering all weld geometry parameters. Using this model and Paris law, a fatigue life of butt weld structures can be estimated.

As a numerical approach, weight function methods require huge calculations, which is time consuming and inconvenient for practical engineering applications, the applications of the FEM to determine crack tip stress field has developed rapidly in recent years. The method has great versatility and enables the analysis of complicated engineering geometry and three-dimensional problems. It also permits the use of elastic-plastic elements to include crack tip plasticity. Basically, two different approaches can be followed in employing finite element procedures to arrive at the required SIF. One approach is the direct method in which K follows from the stress field or from the displacement field around the tip of fatigue cracks [72].

Residual stress intensity factors (K_{res}) are necessary for the fatigue life and crack growth prediction. K_{res} is usually obtained by using weight function solution by applying the residual stress field on the un-cracked component [5].

A general SIF solution to the problem of a crack originating from an angular corner in a finite thickness plate (see Figure 2.13) has been given by Glinka and Niu's weight function [53] and this solution has been widely applied to welded T-plate joints. However, this weight function has only two terms in the approximation and it is only valid for the relative depth (a/t) ≤0.5 [41]. Moreover, O'Dowd et al. [60] mentioned that the range of applicability of the weight function in Ref. [53] is restricted to somewhat limited weld geometries and the T-plate of Figure 2.13(b) falls outside this range. Therefore, they used FEM to determine the weld residual stress intensity factor for cracks of different sizes at weld toe.

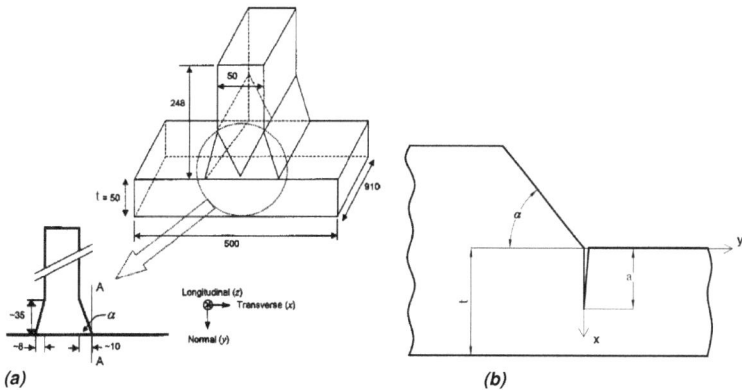

Figure 2.13: a) An edge crack originating from an angular corner in a finite thickness plate [60]; b) geometry of SE702 T-plate weld. All dimensions in mm (not to scale) [41]

Lee et al. [70] determined the residual stress distribution to calculate SIF for T-plate and tubular T-joints using FEA software package ABAQUS.

A method for the determination of weight functions relevant to welded joints and subsequent calculations of SIFs was proposed by Niu and Glinka [53]. The weight function for edge cracks emanating from the weld toe in a T-butt welded joint has been derived by using the Petroski-Achenback crack opening displacement function cited in Niu's work [53]. The weight function makes it possible to study efficiently the effect of weld profile parameters, such as the weld toe radius and weld angle, on SIFs corresponding to different stress systems. They found that the local weld geometric parameters affect the SIFs more than the local stress fields in the weld toe neighborhood.

Guo et al. [41] derived the SIF weight function for a single edge crack originating from the T-plate weld toe from a general weight function form and two reference SIFs. They proposed to apply the weight function to the padded plate geometry (see Figure 2.14). Weight function method is used to calculate SIF because the exact solutions are not always available [41, 53]. Therefore, this method is often used because it enables the SIF for a variety of loading conditions to be calculated by simple integration of the product of weight function $m(x, a)$ and the stress distribution $\sigma(x)$ expression Eq. (2.18) [41].

The coefficients of the weight function are given. The weight function together with the stress distribution on the crack plane obtained from FEA was used to calculate the SIF solutions. The validity of the weight function for the T-plate was verified by the comparison with the numerical data and results from the FEM. Good agreements were achieved. The derived weight function is valid for the relative depth $a/t \leq 0.8$. It is also shown that this weight function is suitable for the SIF calculation for the cracked laser-welded padded plate geometries (see Figure 2.14) under general loading conditions.

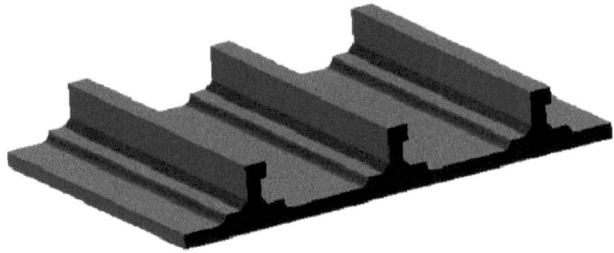

Figure 2.14: The padded plate geometry on fuselage [41]

Applied SIF has been calculated from weight function given by Niu and Glinka. However, this weight function is only valid for the relative depth $a/t \leq 0.5$. Therefore, Guo et al. [41] presented widely applicable weight function solutions with a crack of relative depth (a/t) up to 0.8 originating from the weld toe. They used the $\sigma(x)$ distribution from FEA [41].

More recently Glinka and Shen cited in Refs. [41, 73] have found that the following general weight function expression can be used to approximate weight functions for a variety of geometrical crack configurations subjected to one-dimensional mode-I stress field [41]:

$$m(x,a) = \frac{2}{\sqrt{2\pi\left(1-\frac{x}{a}\right)}} \left[1 + M_1\left(1-\frac{x}{a}\right)^{1/2} + M_2\left(1-\frac{x}{a}\right) + M_3\left(1-\frac{x}{a}\right)^{3/2}\right] \quad (2.19)$$

where, a is the crack length and x the distance along the face of crack. M_i ($i=1, 2, 3$) are parameters which depend only on the geometrical configuration of the cracked body.

In order to determine the weight function $m(x, a)$ for a particular cracked body, it is sufficient to determine the three parameters M_1, M_2, and M_3 in expression (2.19). Because the mathematical form of the weight function (2.19) is the same for all cracks, the same method can be used for the determination of parameters M_1, M_2, and M_3 and calculation of SIFs from Eq. (2.18). The method of finding the M_i parameters was discussed in Ref. [41]. A variety of line load weight functions [72, 74] have been derived and published already. The further details of the M_i determination were discussed in Ref. [41].

$$\begin{aligned} M_1 &= \sqrt{2\pi}(3Y_1 - Y_u) - 4.8 \\ M_2 &= 3 \\ M_3 &= 3\sqrt{2\pi}(Y_u - 2Y_1) + 1.6 \end{aligned} \quad (2.20)$$

where,

$$\begin{aligned} Y_u &= 0.8843 + 4.3274\left(\frac{a}{t}\right) - 39.4056\left(\frac{a}{t}\right)^2 + 284.5721\left(\frac{a}{t}\right)^3 - 1038.1899\left(\frac{a}{t}\right)^4 \\ &+ 2116.4717\left(\frac{a}{t}\right)^5 - 2218.4035\left(\frac{a}{t}\right)^6 + 955.5433\left(\frac{a}{t}\right)^7 \end{aligned} \quad (2.21)$$

$$\begin{aligned} Y_1 &= 0.5854 + 1.8116\left(\frac{a}{t}\right) - 16.4166\left(\frac{a}{t}\right)^2 + 116.5429\left(\frac{a}{t}\right)^3 - 421.5994\left(\frac{a}{t}\right)^4 \\ &+ 848.8765\left(\frac{a}{t}\right)^5 - 876.9786\left(\frac{a}{t}\right)^6 + 370.8612\left(\frac{a}{t}\right)^7 \end{aligned} \quad (2.22)$$

2.9. Conclusions and Discussion for the Current Work

Numerous industrial applications require the consideration of fatigue life of welded joints. It was found that the most of researches were mainly focused on experimental measurements and tests for investigating the fatigue life and the residual stress effects which are time and cost consuming. Other research is based on FEM to evaluate the effect of residual stresses. Little attention has been focused on analytical assessments of the fatigue life calculations of welded joints under the effect of residual stresses, weld geometry, and re-calculation the FAT value of notch cases.

Since weld geometry conditions may differ in various weld joints, traditional empirical relations become invalid in some cases and new models may have to be created for the new local stress distribution and to find the accurate SIF calculations. In line with the traditional da/dN testing approach, nearly all the present da/dN data for welded joints were obtained by using bead removed specimens, for which classical 2-dimensional solutions for SIFs become applicable. It has been reported that welded components with removed bead have different characteristics in fatigue crack growth compared with as-welded components because of the redistribution of welding defects and weld residual stresses (WRS). In most practical engineering applications, removing weld beads thoroughly is impossible or uneconomical especially for heavy welded structures such as those employed on offshore drilling platforms. Using da/dN data obtained from bead removed specimens for fatigue design or fatigue life prediction on as-welded joints may lead to erroneous conclusions [8]. Therefore, the determination of accurate SIF solutions for the correct weld geometry conditions is of practical significance for structural design and fatigue life evaluation of welded structures as carried out in the current work.

Then, the assessment of fatigue life by fracture mechanics method is presented a useful tool to avoid catastrophic failure of any structures and to provide FAT-data.

In spite of considerable fatigue design data which exist for welded joints in the recommendations, the studies of the effects of crack growth parameters and initial crack length are still not clear and have not been discussed enough.

Most researches have presented a limit of ranges for the initial crack length which will affect fatigue life. Therefore, the current work presents the crack lengths and recommendations to use an appropriate length of initial crack.

Different crack lengths were assumed, in addition the activity of the two types of cracks (toe and LOP crack) are considered in some literature, i.e. independent crack growth at the weld toe and weld root was assumed [15]. Then, the lowest FAT values were calculated. In mind of the author, this consideration will conflict with the statistical nature of fracture. This nature states that the normal crack in perpendicular direction to the load will participate only in final failure effectively. Moreover, an initial crack length of 0.2 mm will give a more conservative fatigue life.

In the current work, the gap distance was neglected between two plates to make constancy in results for different geometries which are used. Some other literature used different gap distances for some geometries and neglects for other, and this distance according to the geometry was changed.

In the present work, the SIFs were calculated using the fracture analyses code 2-D program (FRANC2D) due to limited available solutions of SIFs. FRANC2D simulator provides reliable SIF solutions and is more familiar with mesh generation. No 3-D analysis results in welded joints have yet been reported to the author's knowledge.

From the available studies, the effects of weld residual stresses on the variation of fatigue crack growth behavior still remains unclear. However, these effects are critical for fatigue life evaluation of welded joints. In order to evaluate these effects on fatigue life of welded joint structures, SIFs due to residual stresses have to be calculated accurately from the product of weight function and residual stress distributions. In meanwhile, the effects of different residual stress profiles and weld geometries on SIF and fatigue life still need to be explained.

The superposition of K_{res} and K_{app} (i.e., $K_{max,app}+K_{res}$ and $K_{min,app}+K_{res}$) can be used. However, such a superposition will change the resultant R ratio ($R_{res} \neq R$) but the SIF ranges (ΔK) will remain the same ($\Delta K_T = \Delta K_{app}$). Therefore, it is needed to use the fatigue crack growth equation which is sensitive to R like NASGRO equation or add the crack closure phenomenon to the Paris' model or any other model accounting for the state of affairs at the crack tip. Research to date uses the modified Paris' or Forman' law and incorporates them with residual SIF and total SIF to calculate lives which consider only the linear region of crack growth curve. Little studies deal with NASGRO equation to incorporate residual stress effects.

Most studies use the residual stresses measured by neutron diffraction or those calculated from the FEM to calculate the K_{res}.

In mind of the author, up to date no studies were carried out on the numerical calculation of K_{res} for each crack step using the weight function and residual stress distributions that are issued in standards.

According to the above mentioned conclusions there are lacks in some points which will be discussed in the current work. Therefore, the one aim of this book is to calculate the entire crack growth rate with more analytical approach using weight function, polynomial residual stress distributions and calculated initial crack length together with applied SIFs which are calculated firstly from FRANC2D.

Therefore, the following models are used in calculations of the fatigue life. All models have a theoretically-based calculation of the SIF from FE-FRANC2D program as follows:

1. Model-A: Numerical integration of Paris' law using IIW-characteristic parameters (*C*, and *m*). Then, the backward calculation by manipulating the initial crack length was carried out to select the appropriate initial crack length, in turn to determine the FAT values. In this model and due to linear relation between load and SIF, the calculating SIF was scaled with different loads. By using the manipulated a_i, the *S-N* curves and FAT can be obtained for any new geometry.

2. Model-B: Numerical integration of NASGRO equation. The effect of stress ratio, *R*, without the residual stresses were studied using the crack parameters which were calculated from Model-A.

3. Model-C: Using Model-B with the effect of residual stresses. Therefore, the residual stress ratio, residual stress distribution and the residual SIF have to be calculated in this case. The latter was calculated from the product of the residual stresses distribution and the weight function. In turn, different residual stress distributions can be used.

According to the current approaches the new FAT under the effects of geometries and residual stresses can be calculated.

Chapter Three
MODELING OF WELDED JOINTS AND SIF CALCULATION

3.1. Introduction

The stress intensity factor range (SIF) is the inevitable parameter, which must be studied and calculated in fracture mechanics methods. The stress intensity factor, SIF describes the fatigue action at a crack tip in terms of crack propagation. In this work, SIFs have been calculated using fracture analysis code 2-dimensional program, FRANC2D [46].

In welded joints, stress concentrations occur at the weld toe and at the weld root which make these regions the points from which fatigue cracks may initiate [19]. These cracks have been considered in the verification processes using FRANC2D software in this chapter.

Calculating the fatigue life of welded structures and analyzing the progress of these cracks using fracture mechanics technique requires an accurate calculation of the SIF. The existing SIFs were usually derived for one particular geometry and type of loading.

In this study, the finite element method FEM (FRANC2D) was used to calculate the SIF during crack propagation steps. It is verified to be highly accurate, with the direction of crack propagation being predicted by using the maximum normal stress criterion. A developed analytical approach for toe cracks in cruciform welded joints has been used. On the other hand, in case of lack of penetration (LOP) a classical equation from Frank and Fisher [21] is used. The calculated SIF results for some notch cases have been verified with available solutions from International Institute of Welding (IIW) [3, 34], British Standards Institution (BSI) [14, 75, 76] and literature.

A fairly good correlation was obtained and the results have confirmed the use of FRANC2D to simulate different weld geometries. The results are shown and the agreements are pretty good.

3.2. Two Dimensional Analysis of Welded Joints

This chapter describes the 2-dimensional analysis of welded joints using a FE program (FRANC2D). The program was developed by the Cornell Fracture Group from Cornell University, USA [46].

The objectives of this chapter are as follows:
1. Describing the details of welded joint analyses using FE program.
2. Presenting the possibility of two dimensional models to obtain accurate SIF solutions.
3. Validation of the models by studying the effect of mesh size, mesh type and effect of crack increment on SIF values.
4. Studying the effects of joints' geometrical parameters on SIFs.
5. Validation with available solutions from literature (bench marking).
6. Using the validated model to find the SIF for new geometries with insurance of accuracy.

In the application of fracture mechanics to fatigue problems, accurate determinations of the SIF solutions for the case under analysis in the crack propagation phase are needed in order to use the Paris' equation in next steps to calculate the fatigue life.

Table 3.1 shows the cross-section of some welded joints used in this work. Discarding major weld defects, fatigue cracks will originate from the weld toe, and then propagate through the base material, or from the weld root, and then propagate through the weld throat. More details are presented by Hobbacher [3, 32, 34].

Table 3.1: Fatigue cracking of some welded joints that used in this work

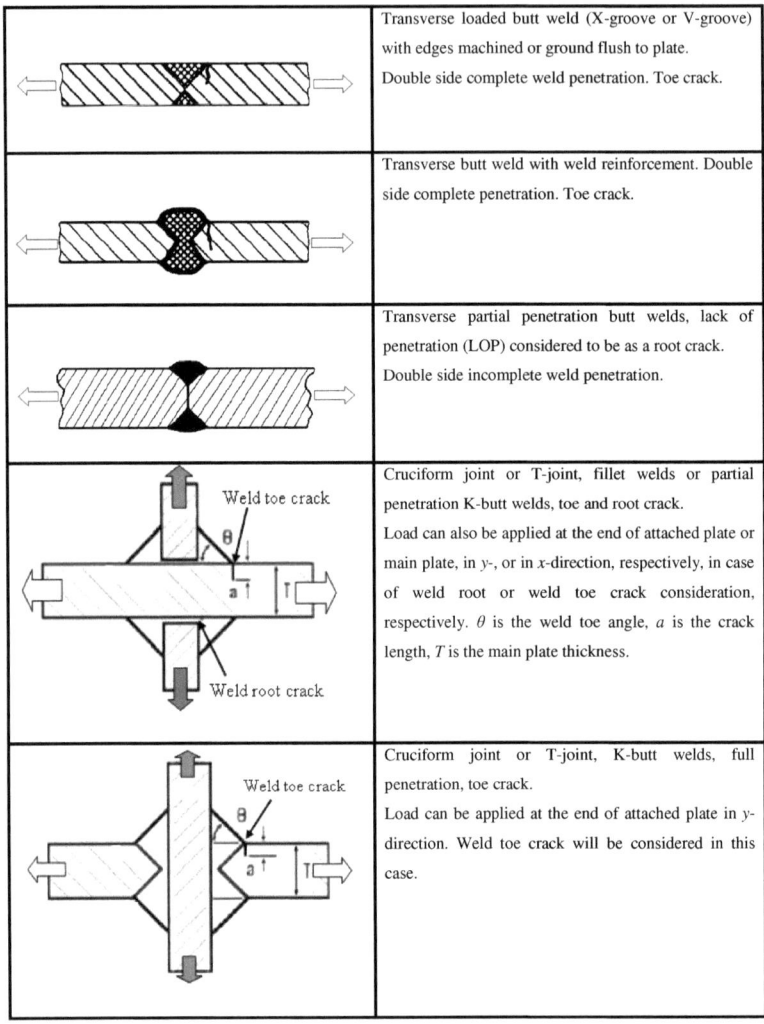

	Transverse loaded butt weld (X-groove or V-groove) with edges machined or ground flush to plate. Double side complete weld penetration. Toe crack.
	Transverse butt weld with weld reinforcement. Double side complete penetration. Toe crack.
	Transverse partial penetration butt welds, lack of penetration (LOP) considered to be as a root crack. Double side incomplete weld penetration.
	Cruciform joint or T-joint, fillet welds or partial penetration K-butt welds, toe and root crack. Load can also be applied at the end of attached plate or main plate, in y-, or in x-direction, respectively, in case of weld root or weld toe crack consideration, respectively. θ is the weld toe angle, a is the crack length, T is the main plate thickness.
	Cruciform joint or T-joint, K-butt welds, full penetration, toe crack. Load can be applied at the end of attached plate in y-direction. Weld toe crack will be considered in this case.

Usually toe cracks have been considered because they are easier to be observed with the naked eye as well as with dye penetration tests and they are often found in many important engineering welded structures. Moreover, there is a high stress concentration located at this point [31, 32].

Shen and Clayton [63] stated that all the cracks were found to be initiated at the weld end toe, in the location of the maximum stress concentration site. SIFs for these cracks have to be determined.

For fracture mechanics treatments, in spite of the fact that several SIF handbooks have been published, it is still difficult to find solutions adequate to many welded configurations [41, 53]. This is mainly due to a wide variety of complex welded geometries and loading systems.

It is known that the SIF for a crack in welded joints depends on the overall geometry of the joint, the weld profile and the type of loading. Thus, the derivation of SIF even for one type of weldment such as a T-butt, cruciform and butt joint requires detailed analysis of several parameters such as plate thickness, weld thickness, weld angle, weld toe radius and loading system.

By characterizing stable macroscopic crack growth using ΔK, it is possible to predict the crack growth rate of a weld under cyclic loading, and hence the number of cycles necessary for a crack to extend from some initial size, i.e., the size of pre-existing crack or crack-like defects, to a maximum permissible size just before catastrophic failures [77].

3.3. Finite Element Analysis

The analysis was undertaken based on the assumption of an initially isotropic elastic material for both the base and its weld metal, in which a crack was subsequently allowed to form and grow according to a fracture criterion. The advantage of symmetry was considered to model quarter or half of the complete geometry.

The FRANC2D is a FE-based simulator for curvilinear crack propagation in planar structures (plane stress, plane strain, and axisymmetric structures). CASCA which is used for the creation the meshes, is a preprocessor for generating initial input files for FRANC2D. FRANC2D program has the ability to analyze a cracked body using special iso-parametric crack tip elements to describe the singularity ahead of the crack tip.

3.4. Mesh Description and Boundary Conditions

Figure 3.1 shows one type of the FE meshes used in the present study, which comprises 8-nodded quadrilateral elements. Also the boundary conditions for T-joint or cruciform joint and for butt joints are shown.

One side of the models was supported in the x-direction or y-direction and a uniform stress distribution was applied at the other side along x-axis or y-axis, respectively. To prevent the model from performing rigid body motions and rotation, one node on the side where the model is supported in the x-direction or y-direction, is also locked in the y-direction or x-direction, respectively, as shown in Figure 3.1.

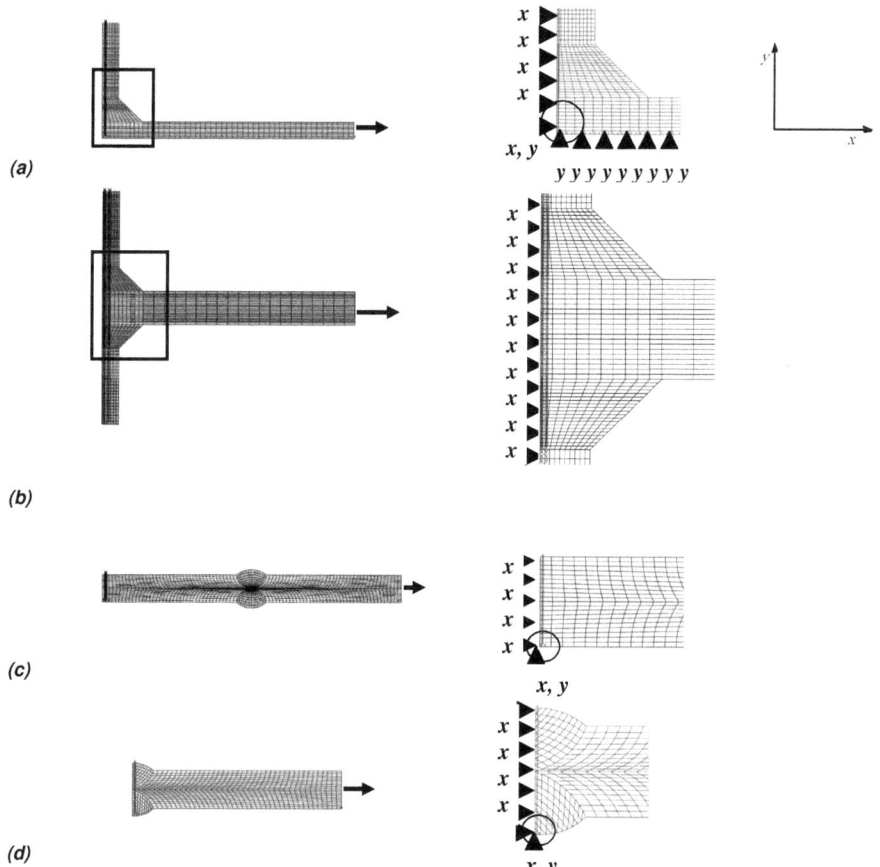

Figure 3.1: Mesh configurations: (a) quarter cruciform model; (b) half cruciform model; (c) complete butt model; (d) half butt model

3.5. Material Properties

The material used in this study for the base material and weld metal was an extra high strength hot rolled steel with the minimum yield strength 550 MPa and the tensile strength minimum 600 MPa and maximum 760 MPa, respectively [31]. Fatigue testing was calculated for an applied loading such that the maximum stress was maintained constant at 200 MPa and 104 MPa for cruciform and butt weld joints, respectively, which fail from the weld toe.

It has to be emphasized in this part, that the input load is distributed with a constant value in tensile direction. Keep in mind that this input load was as force per unit cross section, i.e., in MPa.

Values of Poisson's ratio υ and the modulus of elasticity E were chosen as 0.293 and 210 GPa, respectively.

Experimentally, many structures are optimized by the choice of high strength steel. The reason for this choice is to allow higher stresses and to reduce dimensions taking benefit of the high strength material with respect to the yield criterion. In fracture mechanics, the fatigue strength of a welded joint is not primarily governed by the strength of the base material of the joining members. Hence, the governing parameters in fracture mechanics are mainly the local and global geometry of the joint, i.e., fatigue strength is known to be closely related to the precise geometrical discontinuity of the welded joint [77]. Therefore, the similar material as a single joint between the weld metal and a base metal has been assumed in the simulation. The same steel was used in case of LOP in load-carrying and non-load carrying cruciform and butt joints which fail from LOP with a uniform tensile stress range of 96 MPa and 104 MPa, respectively.

In the absence of specified or measured material parameters, the values given below are recommended to use with fatigue life calculation as characteristic values, C_{char}=3e-13, and m=3 for steel (units in Nmm$^{-3/2}$ and mm).

3.6. Solution Procedure

The determination of the SIF for the 2-dimensional cruciform and butt welded joints has been carried out using LEFM analysis. This method is well encoded in automatic crack propagation in FE analyses FRANC2D. In general, the behavior of crack path propagation (CP) in FRANC2D is almost the same as compared with available experimental results. For every crack length, the values of the SIF were calculated during crack propagation steps by FRANC2D program with suitable boundary conditions, loading, crack growth criteria, and crack direction criteria.

Chapter Three   Modeling of Welded Joints and SIF Calculation

In the following the FE analysis using FRANC2D is described.

3.6.1. Mesh Generation

A mesh generating program CASCA, which is distributed with FRANC2D, was used to create the initial mesh configuration for FRANC2D simulations. Other mesh generating programs can also be used, provided that a translator is available to convert the mesh description to the FRANC2D *.inp format [78, 79]. The procedure for creating a mesh is straightforward, as illustrated in Figure 3.2. To begin with, the problem outline is first created, followed by the division of sub-regions within the problem boundary in Figure 3.2(a). Prior to assigning the type of elements to each of the sub-regions, the boundaries for all sub-regions are divided into the required number of segments, Figure 3.2(b). The resulting mesh for the present simulation is shown in Figure 3.2(c).

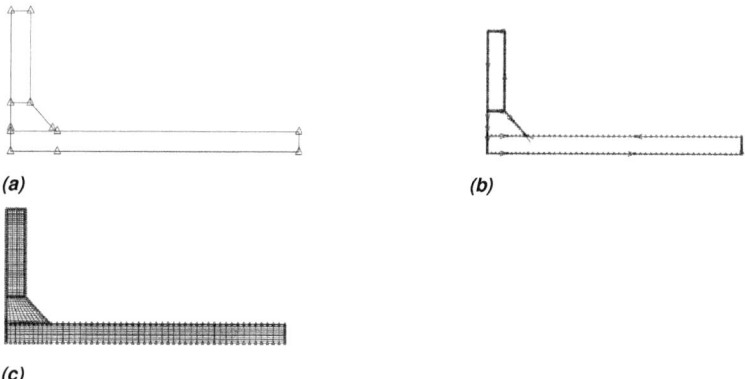

Figure 3.2: Mesh generating procedure for cruciform joint, (CASCA program) [46]: (a) geometry with subdivided areas; (b) subdivided areas into segments; (c) generated mesh

Only one quarter of a cruciform specimen is modeled in FE analysis due to symmetry. The material is assumed to be isotropic linear elastic.

3.6.2. Selection of Material Model

A linear elastic material model coupled to a linear elastic discrete fracture model was used in the analysis. For this analysis and based on fracture mechanics for which linear elastic fracture theory is considered reasonable, it was decided to restrict the analysis of SIF solutions to the linear material behavior. The material assumed to be isotropic linear elastic for base and weld metal.

3.6.3. Crack Propagation

A crack could propagate both from the weld toe and from the root. All calculations of the present work were made on cracks propagating from either the toe or from the weld root separately.

In order to study the capability of the program to simulate crack growth, an initial non-cohesive edge crack was placed at the weld toe of the cruciform and butt weld joints of the weld-base material interface, where it was predicted that critical tensile stresses would occur.

The existence of crack-like imperfections and LOP in the welded joint is normally considered to eliminate the so-called crack initiation stage of fatigue life. Therefore, most of fatigue assessment of welded joints is concentrated on the crack growth stage of the fatigue life. Having specified the location of the crack, the code was able to predict the direction in which the crack would propagate. Prior to performing the analysis, it was necessary to specify the magnitude of crack increment (Δa) and the number of steps over which the crack would propagate. In the present study, a crack increment of 0.5 mm was chosen for all cases. However, this value could be changed according to stability of fatigue life's numerical integration (shown in next chapter). The crack growth was simulated over 10 steps of increment depending on the plate thickness. Moreover, in this study, the crack path was not pre-selected, but crack direction was allowed to change according to the maximum tangential stress criterion [77]. Moreover, the auto-mesh function was carried out automatically.

3.6.4. Modeling Procedures

For every crack length, the values of the SIF factor were calculated during crack propagation steps by FRANC2D program with suitable boundary conditions, loading, crack growth criteria, and crack direction criteria. The results were used to determine the SIF as a function of crack length (a) for a particular case. Then, this function was used to calculate the fatigue life using numerical integration of Paris' equation (see Chapter 4).

The SIF from FRANC2D for weld toe and root cracks were compared with solutions from IIW, BSI and others from literature. A good correlation was obtained, which makes it possible to use FRANC2D to simulate different weld geometries.

The determination of the SIF for 2-dimensional cruciform welded joints has been carried out using linear elastic finite element analysis (LEFE) and the fracture mechanics evaluation. This method is well encoded in automatic crack propagation in FRANC2D:

The following steps have been carried out during the run of FE program (see Figure 3.3) as follows:

1. Determining the input crack length and initiation site.
2. Deleting the region around the crack.
3. Growing the crack to its new location.
4. Setting the minimum number of segments to 1.
5. Placing the rosette around the crack tip.
6. Automatic re-meshing of the region around the crack.
7. Connecting the new mesh region with the former mesh
8. Forward and repeat the steps from the new crack tip location. Therefore, the new directional line will be drawn automatically by the program to indicate the expected next direction for the next crack increment, Δa. Hence, this line seems to be out the boundary as shown in step 8 of Figure 3.3:

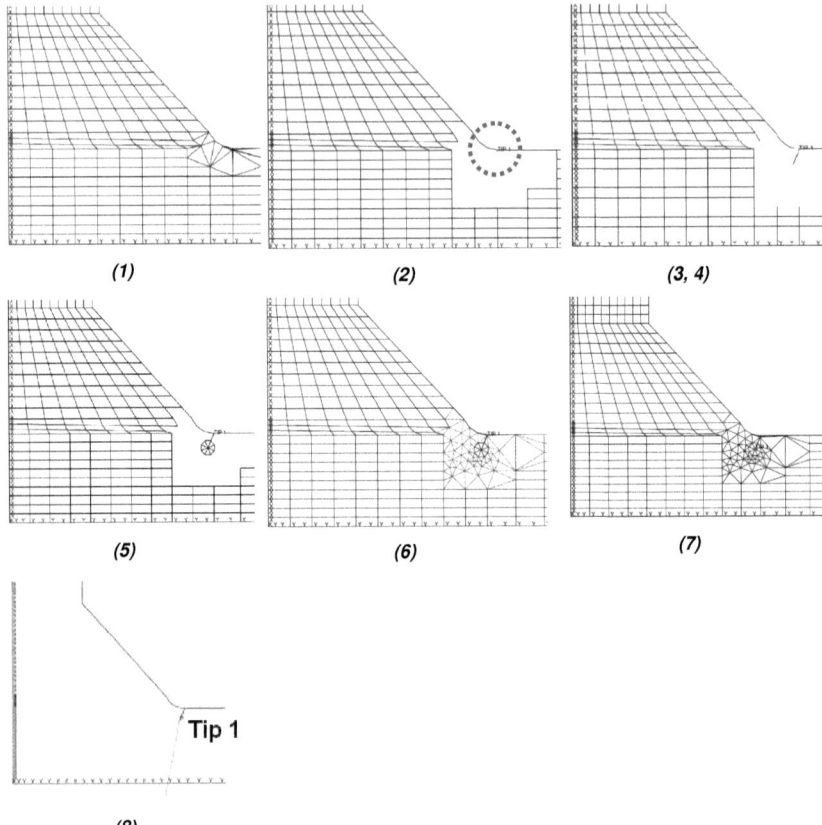

Figure 3.3: Crack growth steps done by FRANC2D

The crack growth with angle of orientation for each step of growing can be calculated using FRANC2D, also the time and cost consuming is lower than for other FEM programs.

The simulation of cracking processes was carried out either automatically or manually as step by step. The former will stop at the final crack according to first crack growth steps and increment. The latter will need to determine the increment for each single step.

The simulation procedures are validated by comparison with experimental and numerical data found in the literature.

3.7. Results Convergence

To investigate the convergence of the results, FE model analyses were performed on models with different meshes and crack increment steps. In addition, the effects of mesh type and symmetry are discussed.

3.7.1. Influence of Mesh Size

Reducing the computational time is the main factor for choosing a coarser mesh in case of bulk geometrical simulations. Karlsson and Lenander [32] stated that the calculation time increases from 10 to 20 times with the finer mesh with little improvements in SIFs calculations, and that justifies the decision to use the coarser mesh in using ANSYS [32].

The choice of mesh size is down to the experience of the user, and unless several iterations of the same model are analyzed, an appropriate mesh density is difficult to define. Therefore, an appropriate mesh size, density, mesh ratio of distribution are recommended to avoid the missing of some elements especially at the corners and edges. Moreover, the insufficient capacity of the memory to solve the larger size specimen with fine meshes will appear. However, a finer mesh in the high stress concentration region is recommended to be used.

Fatigue crack growth simulations of the load-carrying cruciform joint were performed with different levels of mesh refinement (see Figure 3.4) in an attempt to verify that the computed SIFs were mesh independent as shown in Figure 3.5. The dimensions are shown in Figure 3.4(b).

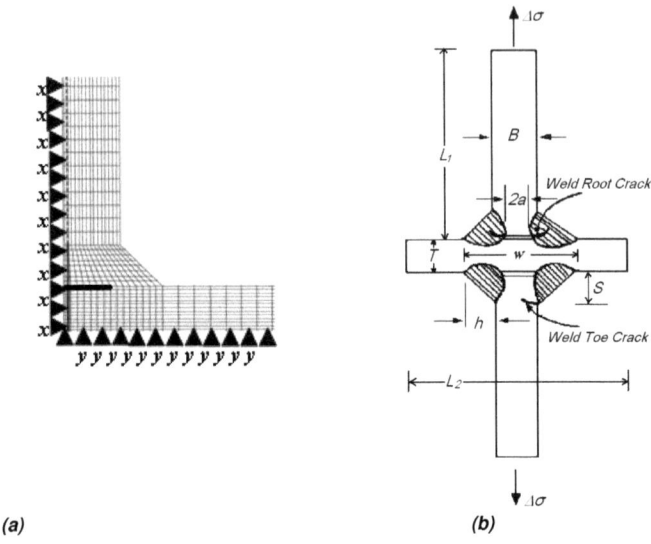

Figure 3.4: *Load-carrying cruciform joint: (a) mesh and boundary conditions in case of LOP (FRANC2D); (b) geometrical parameters*

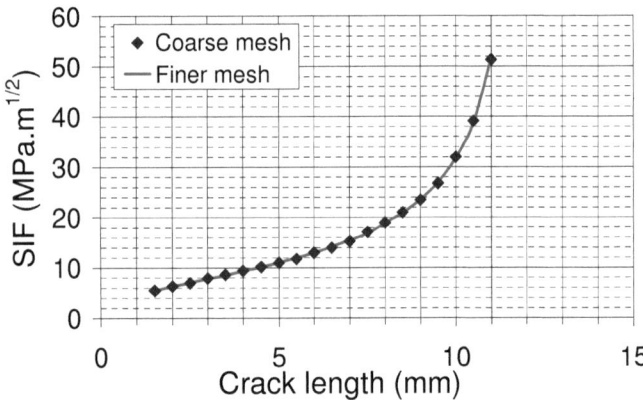

Figure 3.5: *Results convergence for the effect of mesh size and density. $\Delta\sigma=96$ MPa, $a_i=1.5$ mm, $\Delta a=0.5$ mm, $B=T=15$ mm, $h=S=6$ mm*

In different FEM programs, smaller mesh size is recommended to be used near the crack expected region. Increasing of time and badly shaped mesh elements are the main problems reported [32]. Therefore, FRANC2D has advantageous over other FEM programs as shown in Figure 3.6.

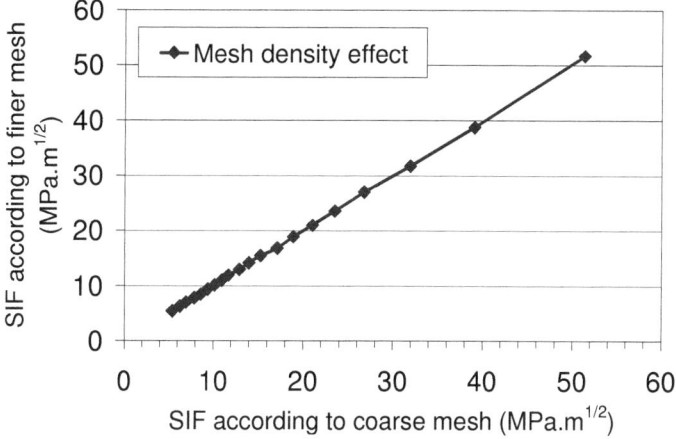

Figure 3.6: SIF results convergence for the effect of mesh size and mesh density. $a_i=1.5$ mm, $\Delta a=0.5$ mm, $\Delta\sigma=96$ MPa, $B=T=15$ mm, $h=S=6$ mm

3.7.2. Influence of Crack Increment

The crack increment Δa is the crack growth step difference $(a_{j+1}-a_j)$, where j is the step's number. The finer and coarse increments were investigated in case of the cruciform joint.

The SIFs were obtained by applying the fracture mechanics method to the FEM. The results of crack length increments effect (Δa) converge as shown in Figure 3.7. There are no influences on the stability of results as shown in Figure 3.8.

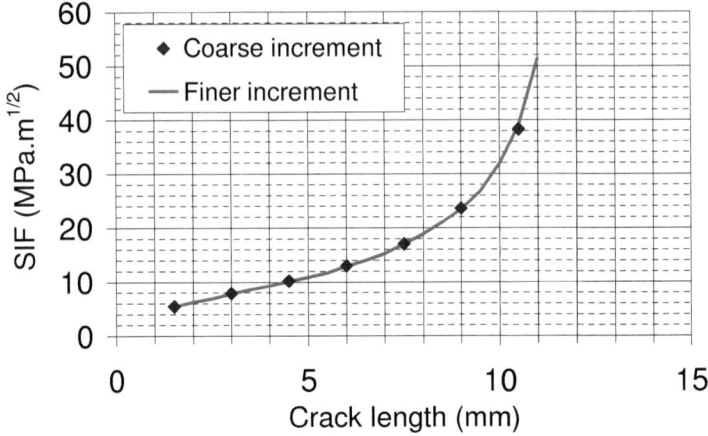

Figure 3.7: Results convergence for the effect of different crack increments with $\Delta\sigma=96$ MPa, $a_i=LOP=1.5$ mm, $\Delta a=0.5$ mm, $B=T=15$ mm, $h=S=6$ mm

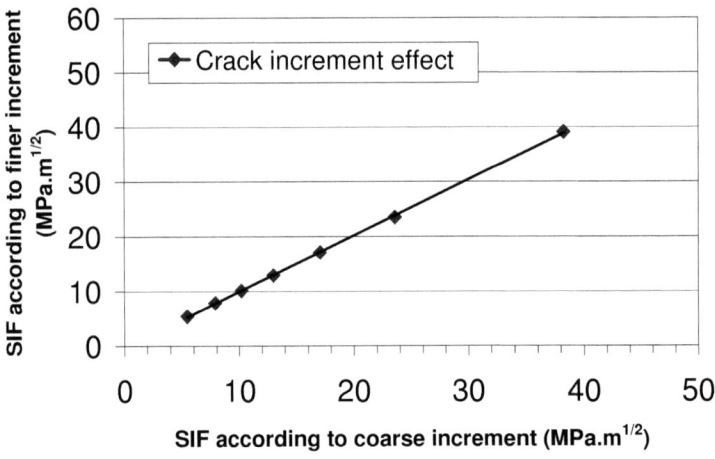

Figure 3.8: SIF results convergence for the effect of different crack increments. $\Delta\sigma=96$ MPa, $a_i=1.5$ mm, $\Delta a=0.5$ mm, $B=T=15$ mm, $h=S=6$ mm

In order to providing confidence in the analysis, the very close agreements between the two types of mesh and increments indicate that these effects on the SIF are negligible. The crack path direction is in agreement with LOP. After some millimeters, the path will be inclined toward the applied load as shown in Figure 3.9.

The simulation procedure is validated by comparing numerical data with experimental results from literature when the LOP is equal to plate thickness (LOP=B, see Figure 3.4b), as shown in Figure 3.9(d-f). In this case, the crack will suddenly curve toward final failure.

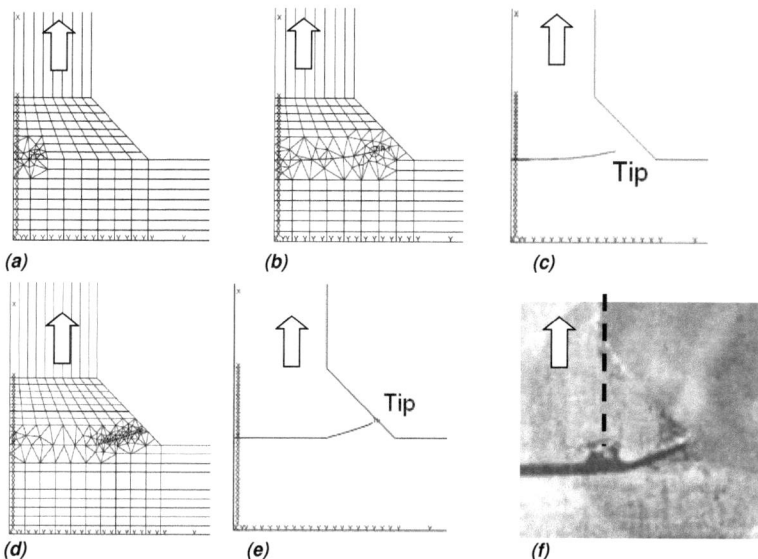

Figure 3.9: *(a) Crack initiation (LOP=B/3); (b) crack growth steps; (c) boundary conditions of root crack joint; (d, e) current simulation (LOP=B); and (f) real propagation path, (LOP=B) [80]*

3.7.3. Influence of Mesh Type

As stated previously, the SIF solutions were not affected by the mesh density or crack increment (Δa size). FRANC2D provides different types of meshes according to the shape and geometry of specimens as shown in Figure 3.10. The comparisons between the Bilinear 4-sided meshes and automatic non-uniform meshes are shown in Figure 3.11 for the weld root and the weld toe crack in cruciform joints. The small crack length is used to have clear effect of the mesh for long crack growth path. The comparison shows quite well agreement. There are no differences in SIFs for using the two different types of meshes.

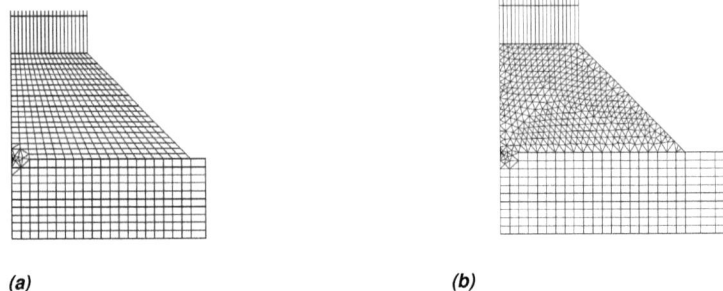

(a)　　　　　　　　　　　　　　　(b)

Figure 3.10: (a) Bilinear 4-sided mesh; (b) automatic mesh

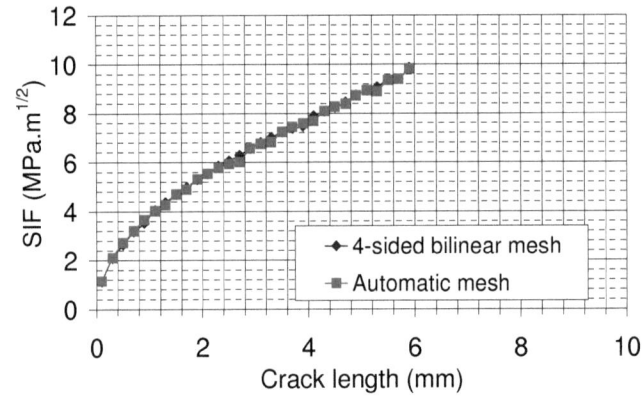

(a)

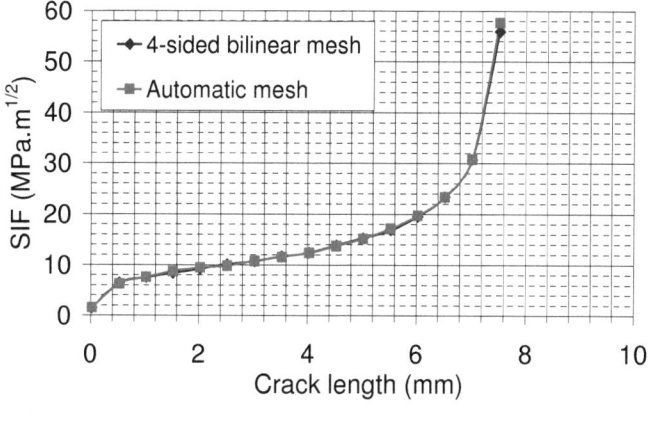

(b)

Figure 3.11: SIFs comparisons between bilinear 4-sided mesh and automatic mesh, $\Delta\sigma=96$ MPa, $B=T=15$ mm, $h=S=10$ mm, $\Delta\sigma=96$ MPa: (a) root crack, $a_i=0.1$ mm, $\Delta a=0.2$ mm; (b) toe crack, $a_i=0.025$ mm, $\Delta a=0.5$ mm

The trend of these curves (3.11a, and b) little bit different than in Figure 3.5 and 3.7 because the using of small crack length and different weld region geometry (h and S) as compared with previous figures for the same crack.

3.7.4. Influence of Symmetry

The advantage of symmetry was taken to model quarter of the complete geometry. Mashiri et al. [25, 81] found that the percentage difference in the number of cycles for fatigue crack propagation life between the simplified half model and a full model is 0.3% using the Boundary Element Analysis System software (BEASY). The complete cruciform joint model with the four crack tips for two parallel cracks (LOPs) and the boundary conditions are shown in Figure 3.12. The FRANC2D FE-based simulation shows that no differences occurred between quarter and complete models as shown in Figure 3.13. Therefore, the quarter models are a reasonable approximation of the cruciform joints and can be used to estimate SIF and the fatigue crack propagation life for these joints.

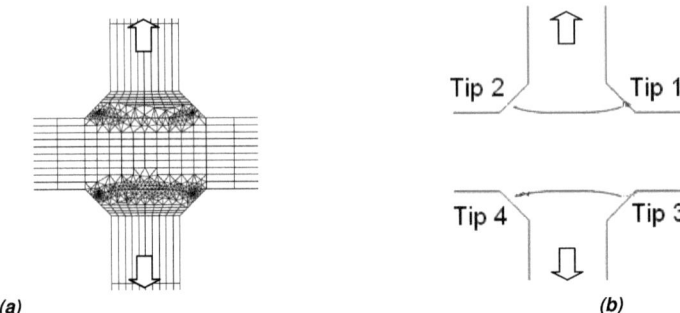

Figure 3.12: *LOP in load-carrying cruciform joints: (a) FE model; (b) crack growth boundary and tips*

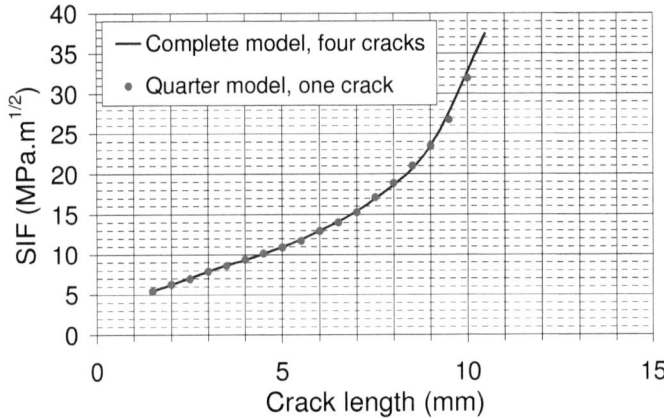

Figure 3.13: *SIFs comparisons between complete model (four crack tips) and quarter model (one crack tip), load-carrying cruciform joint with LOP=1.5 mm, $\Delta a=0.5$ mm, and $\Delta\sigma=96$ MPa, $B=T=15$ mm, $h=S=6$ mm*

If the loading and the cracks are not considered, the cruciform and butt joint are symmetrical about both x- and y-axes. When the crack is present at only one weld toe the model is unsymmetrical.

Quarter symmetry for cruciform joints has been used with 8-nodded quadrilateral elements. Moreover, half of the geometry can be considered as shown in Figure 3.1(a) and (b). Same meshes are used for butt weld joints, as shown in Figure 3.1(c) and (d). The differences in the SIF between the simplified half and quarter geometry of cruciform joints are shown in Figure 3.14. In case of butt weld joint, the comparisons between half and complete geometries are shown in Figure 3.15.

The quarter and half models are therefore reasonable approximations of the cruciform and butt joints failing from the weld toe.

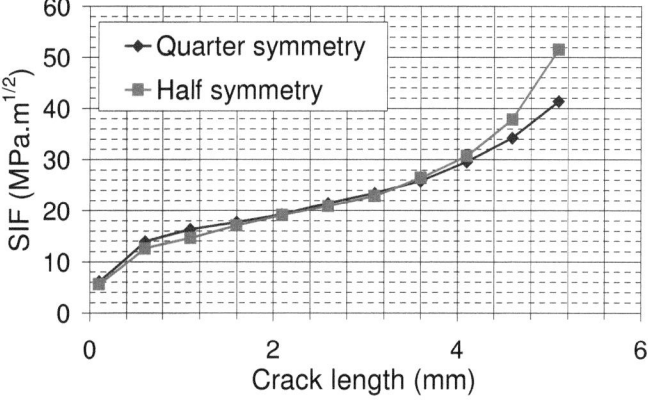

Figure 3.14: *SIFs comparison between half and quarter non-load carrying cruciform joints, toe crack, $\Delta\sigma=200$ MPa, $a_i=0.1$ mm, $\Delta a=0.5$ mm, $B=T=12$ mm, $h=S=8.5$ mm*

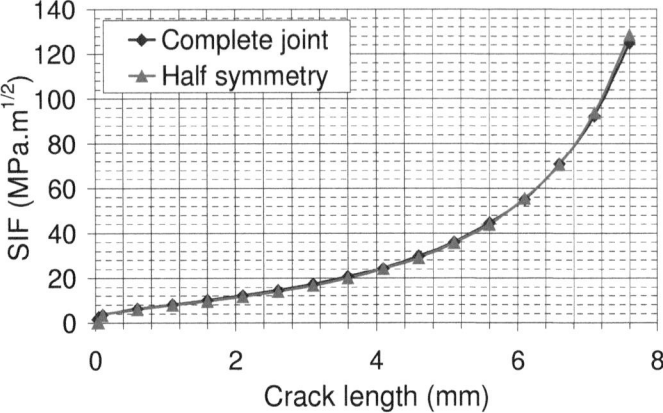

Figure 3.15: SIFs comparison between half and complete transverse butt weld joints, toe crack, $\Delta\sigma=104$ MPa, $a_i=0.1$ mm, $\Delta a=0.5$ mm, thickness=10 mm

3.8. Selection of the Notch Cases

Typical notch cases which are used in this work are presented in Table 3.1. In parallel joints toe cracks and LOP are frequently encountered defects.

Toe cracks have been simulated in case of non-load carrying welded joints, namely cruciform and butt welded joints where the weld metals have been assumed as complete penetration. The root cracks are found in load-carrying joints.

The most conventional joints in engineering structures are butt weld and cruciform fillet welded joints. According to the crack type, location and applied load position, these joints can be classified into load-carrying and non-load carrying joints. In the latter, the fatigue cracks usually occur at the weld toe, where the load is applied along the x-direction. By contrast, in the former, cracks starts from LOP where the load is applied along the y-direction. Due to symmetry, the quarter modelled joint can be used. Figure 3.16 shows the used FE models and the sites of cracking. The high stresses are located at weld toe transition and in addition at the crack tip of LOP. This explains the reason for crack propagation from these locations.

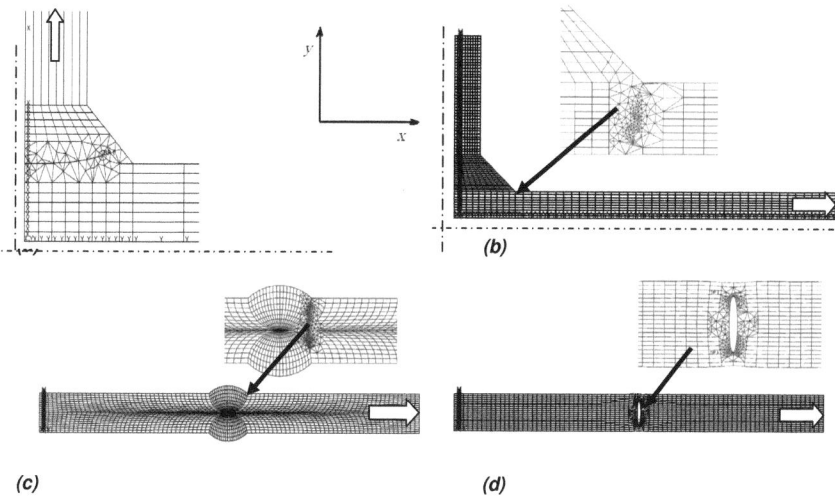

Figure 3.16: FE modelling, all cases have a transition radius of 1 mm: (a) root crack in load-carrying cruciform joint; (b) toe crack in non-load carrying cruciform joint; (c) toe crack in butt joint; (d) LOP in butt joint

3.9. Stress Intensity Factor Calculations

The main expression for the SIF is:

$$K_I = \Delta\sigma \, (\pi a)^{1/2} f(b) \tag{3.1}$$

The applied stress is $\Delta\sigma$ and $f(b)$ is the correction factor. The stress intensity correction factor $f(b)$ is a function of the crack length to thickness ratio $f(a/t)$. The analytical expression for $f(b)$ can be of the tangent or secant type. It is not too obvious how to choose between the tangent and secant expression for estimating the value of parameter $f(b)$, although the "secant" version appears to be a bit more straightforward [82].

$$f(b) = \left(\sec\frac{\pi a}{2t}\right)^{1/2} \tag{3.2}$$

where the term $(\pi a/2t)$ is given in radians.

Some modifications have been carried out on geometrical correction factor as shown for joints in this chapter. In this work, FRANC2D was used to calculate the opening mode SIF using fracture mechanics approach where the stress ranges were fully effective (i.e., the crack remains open during the propagation period where the initiation time is discarded). The influence of K_I on fatigue crack growth was based on the maximum tangential stress criterion by Erdogan and Sih [83]. This criterion assumes that the predicted propagation path of the fatigue crack is perpendicular to the maximum principal stress and the crack grows under opening mode and stress ranges were fully effective [15].

3.9.1. SIF of Load-Carrying Cruciform Joints

For fillet welds and partial penetration welds, the unpenetrated region may be considered to act as an initial crack (see Figure 3.16a). The SIF resulting from the un-penetrated region will depend on the detailed geometry and hence a significant effect on fatigue behavior is to be expected.

Frank and Fisher [21] derived an empirical formula for obtaining the SIF of fatigue cracks that originate from weld roots based on the results of FEA. They performed a fatigue crack propagation analysis using their empirical formula and proposed a stress range calculation formula to evaluate fatigue strength. However, their study examined only joints with isosceles-triangle-shaped fillet welds [49, 84]. The polynomial expression for SIF range (ΔK), for a crack at the weld root in case of isosceles weld shape of a load-carrying cruciform joint developed by Frank and Fisher is given below:

$$\Delta K = \frac{\Delta \sigma}{1+2(h/B)} [A_1 + A_2 a/w][\pi a Sec(\pi a/2w)]^{1/2} \qquad (3.3)$$

where h and S are the fillet weld leg length on the main and the attached plate, respectively, and h/B is defined as the weld size. T and B are the main and the attachment plate thickness, respectively. a is half the crack length ($2a$=LOP) and $\Delta \sigma$

is the applied tensile stress to attachment plate. All the parameters are shown in Figure 3.17. The width *w* of the fillet weld as shown in Figure 3.17 is:

$$w = B + 2h \tag{3.4}$$

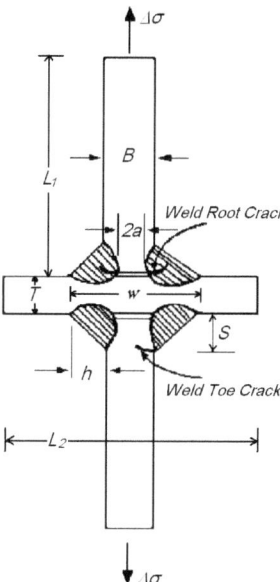

Figure 3.17: *Load-carrying cruciform fillet welded joint geometry*

A_1 and A_2 are functions of weld size (h/B) given by:

$$A_1 = 0.528 + 3.286(h/B) - 4.361(h/B)^2 + 3.696(h/B)^3 - 1.874(h/B)^4 + 0.415(h/B)^5 \tag{3.5}$$

$$A_2 = 0.218 + 2.7717(h/B) - 10.171(h/B)^2 + 13.122(h/B)^3 - 7.775(h/B)^4 + 1.785(h/B)^5 \tag{3.6}$$

The International Institute of Welding (IIW) adopted Frank, Fisher formula's for the SIF which is valid for h/B from 0.2 to 1.2 and for (a/w) from 0.0 to 0.7 [21]. This formula was then improved by British Standards Institution PD6493 [85] and by BSI7910 [86], where the range of the SIF, ΔK, can be written as:

$$\Delta K_I = M_k \Delta \sigma \sqrt{\pi a \operatorname{Sec} \frac{\pi a}{w}} \tag{3.7}$$

The parameter K_I is known as the opening mode-I stress intensity and represents the strength of the stress surrounding the crack tip. M_k is the stress concentration magnification factor which is defined as the ratio of the SIF of a cracked plate with a stress concentration to the SIF with the same cracked plate without the stress concentration [77].

$$M_k = A_0 + A_1 \left(\frac{2a}{w}\right) + A_2 \left(\frac{2a}{w}\right)^2 \tag{3.8}$$

The values of A_0, A_1 and A_2 for equal leg length fillet welds and unity ratio of main plate to attachment plate thickness are as follows [19]:

$$A_0 = 0.956 - 0.343 \left(\frac{h}{B}\right) \tag{3.9}$$

$$A_1 = -1.219 + 6.210 \left(\frac{h}{B}\right) - 12.220 \left(\frac{h}{B}\right)^2 + 9.704 \left(\frac{h}{B}\right)^3 - 2.741 \left(\frac{h}{B}\right)^4 \tag{3.10}$$

$$A_2 = 1.954 - 7.938 \left(\frac{h}{B}\right) + 13.299 \left(\frac{h}{B}\right)^2 - 9.541 \left(\frac{h}{B}\right)^3 + 2.513 \left(\frac{h}{B}\right)^4 \tag{3.11}$$

The limits of validity for this formula are within the following range:

$$\frac{B}{T} = 1, \ 0.2 \langle \frac{h}{B} \langle 1.2 \ , \text{ and, } 0.1 \langle \frac{2a}{w} \langle 0.7$$

In all above formulas, the unity leg lengths have been assumed ($h=S$). Motarjemi et al. [19] calculated numerically the SIFs for cruciform and T-steel welded joints using ABAQUS. SIF results of the cruciform joint were firstly compared with the above formula for the case of equal attachment and main plate thickness ratio ($B/T=$ 1). Results were then calculated for welded joints with $B/T < 1$ where the above formula is no longer valid. No analysis was carried out for $B/T > 1$ as this is not a recommended configuration.

The investigations of the effects of other weld parameters, such as weld toe length on both plates are indistinct. These effects have not been studied and need therefore to be investigated.

3.9.2. SIF of Non-Load-Carrying Cruciform Joints

The cruciform weld joint with equal attachment and main plate thickness ($B/T=1$) is shown in Figure 3.18. The non-load carrying cruciform joint is made from the welding of stiffener plate (attachment plate) perpendicular to the main plate (loading plate). Sheet thickness was taken as 12 mm. The advantage of symmetry was taken to model quarter of the complete geometry. Here, a is the initial crack length located at weld toe, h is the weld leg length over main plate side and $w=2h+B$.

For an elliptical crack at the toe of a fillet welded joint, the range of the SIF can be written as [26]:

$$\Delta K_I = \frac{M_k Y_u}{\phi_0} \Delta\sigma \sqrt{a} \tag{3.12}$$

In Eq. (3.12) M_k is the stress concentration magnification factor, $\Delta\sigma$ is the nominal tensile stress range applied on the main plate and a is the crack depth [77].

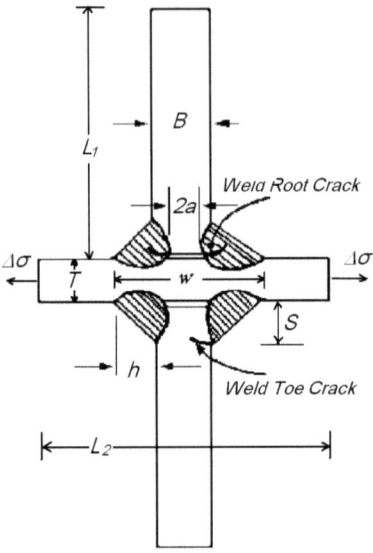

Figure 3.18: Non-load carrying cruciform fillet welded joint geometry

ϕ_0 is the complete elliptical integral defined as [26]:

$$\phi_0 = \int_0^{\pi/2} \left[1 - \left(1 - \frac{a^2}{c^2}\right)\sin^2\phi\right]^{1/2} d\phi \qquad (3.13)$$

where ϕ is parametric angle of ellipse.

In this analysis the M_k-factor functions are based on **continuous** edge cracks. Hence, the crack aspect ratio is zero, $a/2c = 0$, and $\phi_0 = 1$ [77]. The correction term Y_u for a double-edge crack in a plate under tensile loading, Eq. (3.14), given by Brown and Srawley [87] was applied:

$$Y_u = 1.98 + 0.36\left(\frac{2a}{T}\right) - 2.12\left(\frac{2a}{T}\right)^2 + 3.42\left(\frac{2a}{T}\right)^3, \quad 0 \langle \frac{2a}{T} \langle 0.95 \qquad (3.14)$$

Chapter Three   Modeling of Welded Joints and SIF Calculation

T is the main plate thickness. For fracture mechanics treatments Maddox [26] introduced an alternative concept of M_k as a magnification of the SIF, which would be present for a crack of the same geometry but without the presence of the weld. Further work on M_k values for cracks at weld toes has been carried out by Lie [88], Thurlbeck [89] and Bowness and Lee [90].

IIW introduced a systematic set of formulas for M_k values for different welded joints [3, 32, 34]:

$$M_k = C_1 \left(\frac{a}{T}\right)^k, M_k \leq 1 \tag{3.15}$$

$$C_1 = 0.8068 - 0.1554 \left(\frac{S}{T}\right) + 0.0429 \left(\frac{S}{T}\right)^2 + 0.0794 \left(\frac{h}{T}\right) \tag{3.16}$$

$$k = -0.1993 - 0.1839 \left(\frac{S}{T}\right) + 0.0495 \left(\frac{S}{T}\right)^2 + 0.0815 \left(\frac{h}{T}\right) \tag{3.17}$$

where h and S are the weld leg length on main and attachment plate side, respectively, T and B are the main and attached plate thickness respectively and a is the initial crack length from weld toe. By substituting the solution of M_k from Eq. (3.15) and the correction term from Eq. (3.14) in Eq. (3.12), the SIF can be calculated for cruciform joint fails from the weld toe, as shown in section 3.10.4.1.

3.9.3. SIF of Transverse Butt Weld Joints having Toe Crack

Butt weld joints with the thickness t equal to 10 mm have been simulated. The weld bead height (H) and the weld bead width (W) were 2 and 10 mm, respectively. Here, ρ is the weld toe radius, θ ($\theta=180-\gamma$) is the weld toe angle, see Figure 3.19.

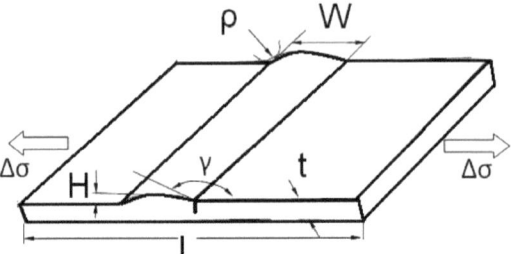

Figure 3.19: Weld geometrical parameters at transverse butt weld joint, toe crack

For the validation of the prediction results, the range of SIF for the welded specimens was calculated using the following empirical Eq. (3.18) [91]:

$$\Delta K = \Delta\sigma\sqrt{\pi a}\left(1.12 - 0.23(a/t) + 10.55(a/t)^2 - 21.72(a/t)^3 + 30.39(a/t)^4\right)$$

(3.18)

where $\Delta\sigma$ is the stress range, a is the crack length and t is the thickness of the plate. The results from Eq. (3.18) were compared with those obtained using FRANC2D.

3.9.4. SIF of Transverse Butt Weld joints having LOP

Insufficient accesses of weld metal in groove size of butt weld joint will create a LOP defect which is considered to be the source of crack propagation. Figure 3.20 shows this defect. Traditional path of crack is normal to the applied load. IIW listed this type of notch case under class 217 as incomplete weld penetration. The simulation of flush welded in both sides is not important.

Theses aspects will be discussed in case of toe cracks because the flush weld bead height will cause decline in fatigue strength (see Chapter Four). In the mind of the author, the weld bead geometries have no effect on crack path and SIF calculations in case of LOP. In the literature, few studies for this type of defect in butt joints have been presented. This might be due to the consideration of this configuration as unacceptable from the design point of view. The root side of the weld is frequently neglected in the design process as pointed out in Ref. [36] and no theoretical solutions are monitored until now.

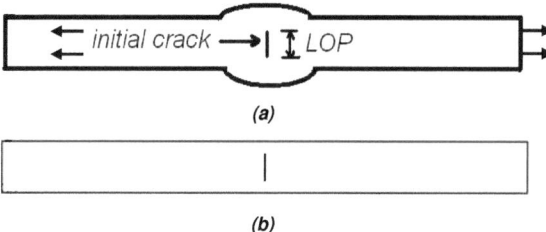

Figure 3.20: *(a) Transverse butt weld joints with LOP; (b) FE model without overfill*

3.10. Results and Discussion

The crack growth simulation and SIF calculation are presented in the following sections for cruciform and butt welded joints.

3.10.1. Cruciform Fillet Weld Model

Particularly, there are competing sites for initiation of fracture at the weld toe and at the weld root for fillet weld joints. At these points, a high stress concentration occurs which makes these regions the easier points from which fatigue cracks may initiate [19]. It is important to recognize that for both fracture and fatigue, the cracks may initiate and grow in the vicinity of the welds during service life even if the applied stresses are well below the yield limit.

3.10.2. Stress Distribution

Figure 3.21 shows the stress near the weld toe is equal to 364 MPa whereas the applied load is 200 MPa.

The maximum and minimum principal stress distributions for uniform load distributions of cruciform without crack are shown in Figure 3.21 for uncracked edge. The local stress will increase with existence of a crack because of the local stress field at the crack tip of 0.1 mm will be more severe. Figure 3.22 gives the principal stress distributions in case of presence of an edge crack.

A maximum tangential stress criterion was used to predict the crack growth direction under mixed mode K_I-K_{II} conditions, so that according to this criterion, the crack path direction will change automatically and was not pre-selected. In the FRANC2D, the site and curved crack growth path of continuous toe cracks were taken into account as shown in Figures 3.23 and 3.24, respectively. Normally this kind of non-load carrying welded attachments always fails from the weld toe as shown in Figure 3.25.

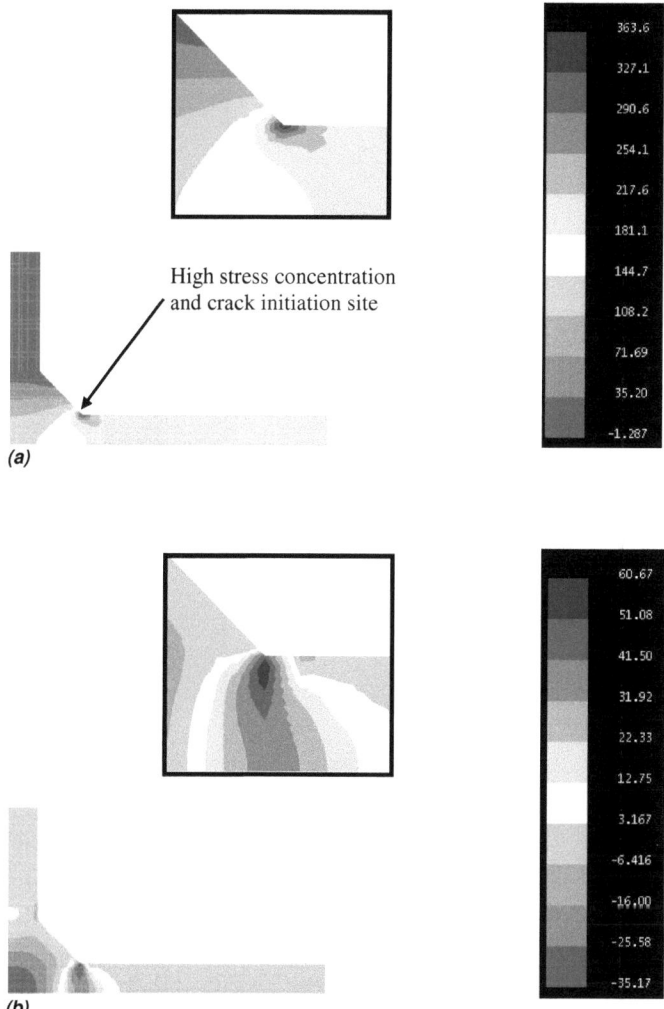

Figure 3.21: (a) and (b) Maximum and minimum principal stress distribution for cruciform joint (MPa) without toe crack, respectively. The stress at weld toe is 364 MPa for applied load equal 200 MPa

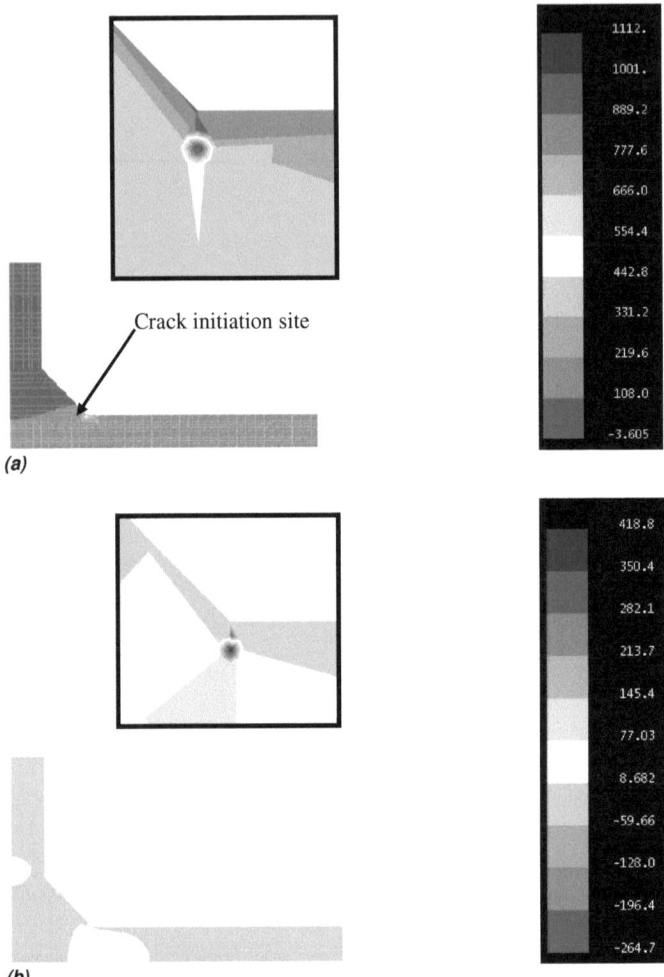

Figure 3.22: (a) and (b) Maximum and minimum principal stress distribution (MPa) in existence of edge toe crack ($a_i=0.1$ mm), respectively. The local stress at weld toe was exceeded due to local edge crack

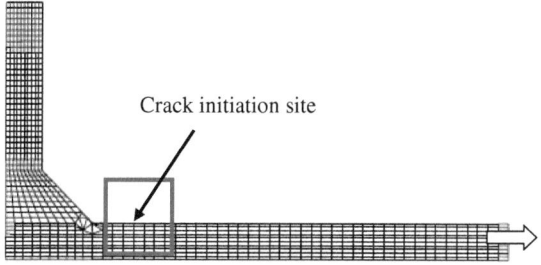

Figure 3.23: Crack formation at weld toe due to local geometry effects, $a_i = 0.1$ mm

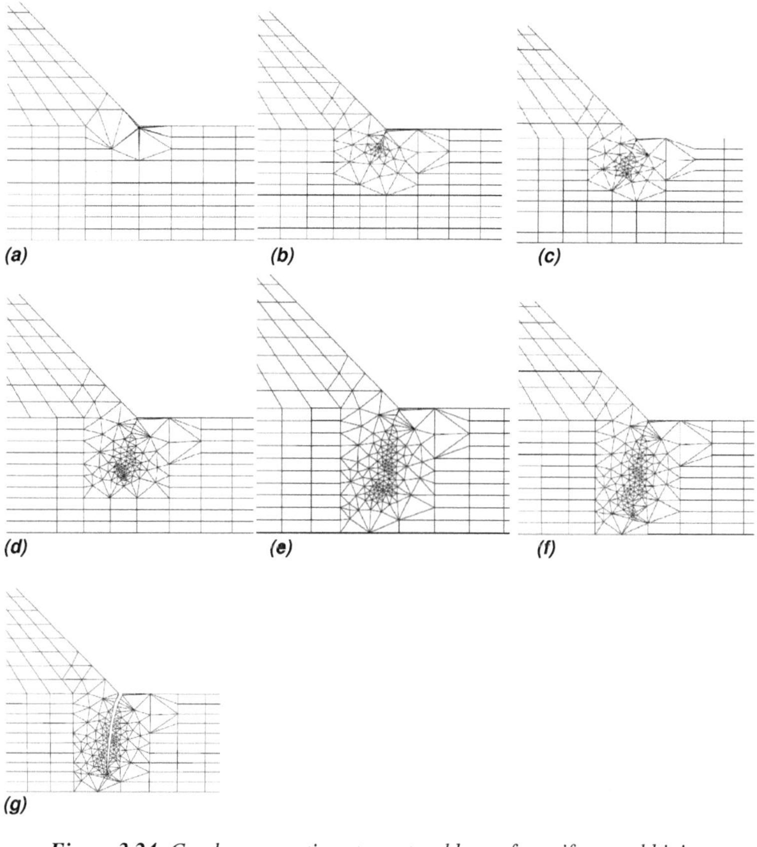

(a) (b) (c)

(d) (e) (f)

(g)

Figure 3.24: Crack propagation steps at weld toe of cruciform weld joints

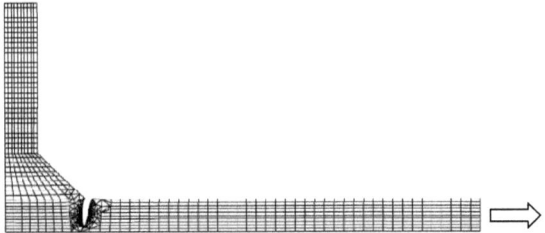

Figure 3.25: Crack deformation of cruciform weld joint

3.10.3. Butt Weld Model

The same steel was used for welding the butt welded joints. Weld toe crack initiated and grows through the sheet thickness of 10 mm.

Figure 3.1(c, and d) shows the modelling and boundary conditions of butt weld joint that are used to calculate the SIF. Figures 3.26 and 3.27 show the crack initiation and propagation, respectively. The final crack deformation is shown in Figure 3.28.

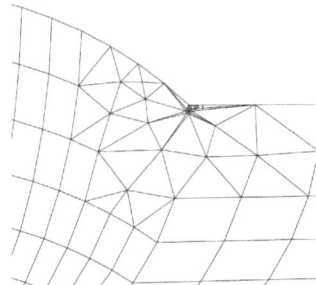

Figure 3.26: Screenshot from FRANC2D crack tip at weld toe, $a_i = 0.1$ mm

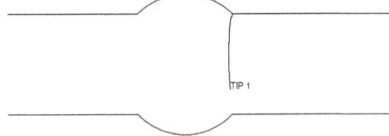

Figure 3.27: Screenshot from FRANC2D of the crack growth direction, toe crack

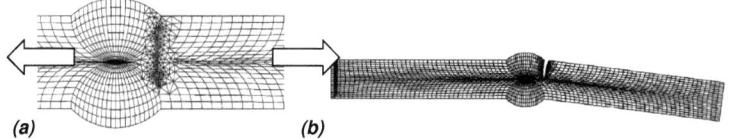

Figure 3.28: Screenshot from FRANC2D of the crack growth direction with mesh: (a) final crack propagation; (b) crack deformation and failure of butt weld joint

3.10.4. Results of Verifications

Cruciform joints and butt weld joints are the most conventional weldments which are used in engineering structures. Accurate predictions of crack shape changes and fatigue lives require accurate SIF estimates because of the power-law nature of the fatigue life crack growth law. It has to be emphasized that all cases have a transition radius of 1 mm, as suggested by different investigations [32].

3.10.4.1. SIF of Non-Load-Carrying Cruciform Joints

Existing SIF solutions are mostly derived from 2-D plane stain models containing edge cracks [8].

The results of the analytical solution for the cruciform joints with equal main (web) and attached (flange) plate thickness are plotted in Figure 3.29, based on the correction terms Eq. (3.14), and M_K, Eq. (3.15) from IIW together in Eq. (3.12), see section 3.9.2. This solution is compared with the direct calculations of SIF as a function of crack length from FRANC2D. It can be seen that the developed approach is consistent with the solution from FE for cruciform joint failure from the weld toe as shown in Figure 3.30.

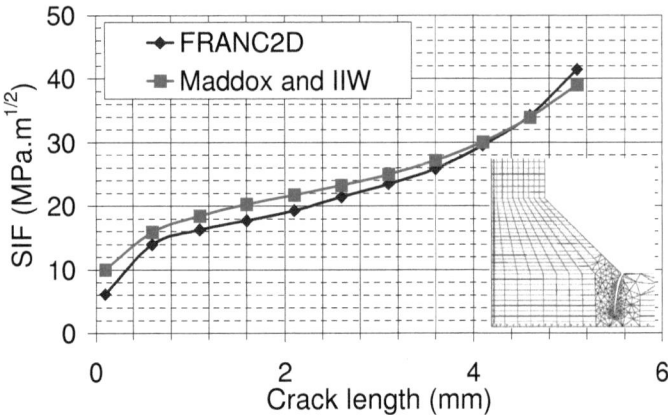

Figure 3.29: SIF as a function of crack length for cruciform failure from weld toe joint as compared with FEM, $\Delta\sigma=200$ MPa, $B=T=12$ mm, $h=S=8.5$ mm, $a_i=0.1$ mm, $\Delta a=0.5$ mm

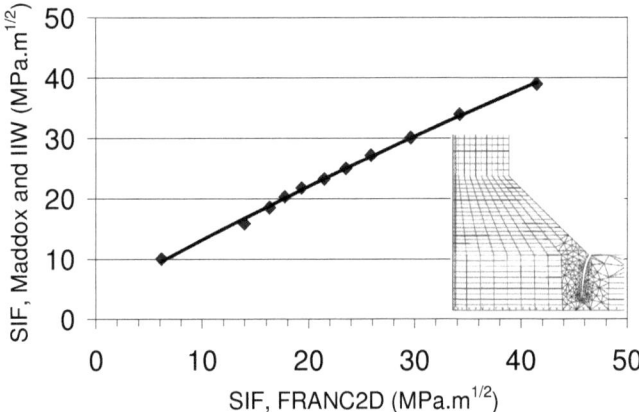

Figure 3.30: SIF values for cruciform joints failure from the weld toe calculated using FRANC2D compared with developed analytical approach calculation (Eq. 3.12). $\Delta\sigma=200$ MPa, $B=T=12$ mm, $h=S=8.5$ mm, $a_i=0.1$ mm, $\Delta a=0.5$ mm

According to Ref. [77], the correction term Y_u was given for double edge crack in a plate under tensile loading. Therefore, the polynomial form of this term (Eq. 3.14) will make a difference in SIF solution (Eq. 3.12) as compared with current FE solution (FRANC2D). Moreover, Eq. (3.12) was developed for semi-elliptical surface crack in a plate under tension as presented in Refs. [26, 77].

FRANC2D has predicted the solution for edge crack at the weld toe of cruciform fillet weld joint. This single edge crack has a different crack path according to the assumption of maximum stress criteria. The crack has inclined at first with respect to the applied load and then turned to direction normal to the applied load (see Figure 3.24). Therefore, the differences started at the beginning of crack propagation and after some millimeters, there are some consistencies with analytical solution and the two solutions become closer for larger crack depths (see Figure 3.29). The reason is that the crack path will become similar to the behavior of double edge crack in a plate under tensile loading.

3.10.4.2. SIF of Load-Carrying Cruciform Joints

From literature [16, 17, 18] it is evident that most of the investigations on fatigue life prediction of fillet welded joints are based on the weld toe failure. Some other studies have considered the fatigue behavior of fillet welded cruciform joints failing from the root region.

In studies [19, 20, 21, 22, 54, 84, 92] the fatigue behavior and SIF of cruciform and T-welded joints of steels failing from the root (LOP) have been studied.

The maximum stress intensities develop when the plane of the crack path is normal to the direction of the primary tensile stress. The SIFs for cruciform joints with LOP were calculated using FRANC2D and were compared with solutions from Frank and Fisher [21], BSI 7910 [14, 85] and IIW [3, 34] and are compared in Figure 3.31.

To study the effect of curved crack path and convergence of results, two FE solutions have been used. In the first solution, the SIFs are calculated only for initial crack step, i.e., a_i= 1.5, 2, 5, 7.5, 9, 10, and 11.5 mm, where the crack growth plane still straight forward. In the second solution, the SIF was calculated, when the initial LOP crack is allowed to grow to its final length (from $2a_i$=LOP to failure) i.e., the crack growth plane is curved. The comparison of both solutions in Figure 3.31 shows that the solutions from BSI 7910 are more realistic and have quite good agreement with FEA, more than IIW solution. The latter solution is worse because the non-modified SIFs equation was used.

The first FRANC2D calculation is more consistent with that from BSI up to the finally calculated crack length, however, the first FRANC solution assumed a straight line crack path normal to the applied load.

The second calculation from FRANC shows a good fit with BSI up to a crack length equal to the plate thickness ($a=B/2$) where the crack is normal to the applied load and has a straight path. Over this length, the comparison has a little bit diverged because the crack will be inclined with respect to the original crack path.

Since the author found negligible variation between the two FRANC2D solutions and the modified SIF solution from BSI 7910, it is proposed to use FE for calculating the SIF range in various shapes failing from the root gap.

The most important results are that the solutions from BSI and IIW have ignored the real effect of crack path inclination as shown in comparison between the second FRANC solution and those from BSI and IIW in Figure 3.31.

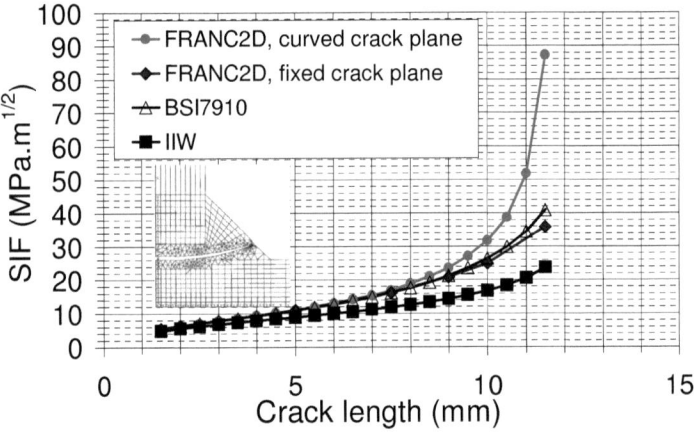

Figure 3.31: *Comparisons of different SIF solutions, $\Delta\sigma=96$ MPa, $B=T=15$ mm, $h=S=6$ mm, $a_i=1.5$ mm, $\Delta a=0.5$ mm*

The above comparisons may lead to differences in fatigue life according to which SIF solutions are used. IIW presents a lower value of SIF. BSI and FRANC2D show a good coalescence in SIF solutions. The fatigue life comparison between these SIF solutions will be presented in Chapter 4 to show their effect on fatigue life.

3.10.4.3. SIF of Transverse Butt Weld Joints having Toe Crack

The comparison of results obtained for the butt weld is shown in Figure 3.32. It can be seen that the results from FEM is close to the results obtained from Eq. (3.18). FRANC2D compares well with empirical equations and appears to be more realistic because different deviations from ideal geometries can be realised.

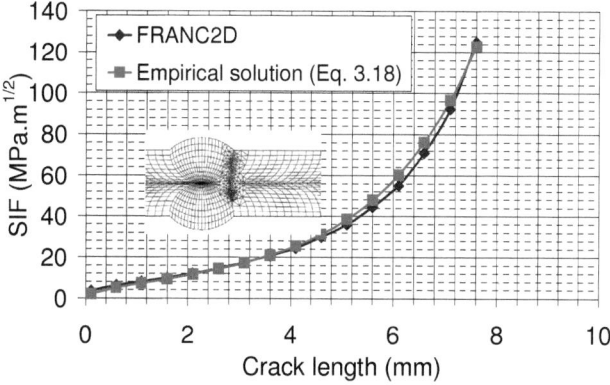

Figure 3.32: FRANC2D and empirical solutions of SIF for butt weld joint according to Broek [91], $\Delta\sigma=104$ MPa, $a_i=0.1$ mm, $\Delta a=0.5$ mm, thickness=10 mm

3.10.4.4. SIF of Transverse Butt Weld Joints having LOP

The crack is simply propagating in direction normal to the applied load due to the nature of these defects and this butt joint. The SIF traditionally increases as the crack length increases as shown in Figure 3.33. There are no effects of weld bead dimensions on the SIF results for this type of joints, i.e., the effect of weld bead height and width.

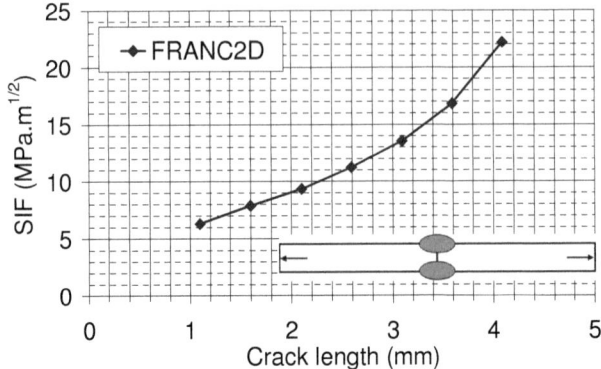

Figure 3.33: SIF of transverse butt weld joint having LOP, $\Delta\sigma=104$ MPa, $a_i=LOP=1.1$ mm, $\Delta a=0.5$ mm, thickness=10 mm

In the present work, the poylnomial equation could be useful to develope a geometrical function in case of LOP, as follows:

$$f(a/t) = \left(55.483(a/t)^4 - 56.828(a/t)^3 + 26.002(a/t)^2 - 6.0102(a/t) + 0.9679\right) \quad (3.19)$$

Then the SIF range is:

$$\Delta K = \Delta\sigma\sqrt{\pi a}\left(55.483(a/t)^4 - 56.828(a/t)^3 + 26.002(a/t)^2 - 6.0102(a/t) + 0.9679\right) \quad (3.20)$$

where $\Delta\sigma$ is the stress range, a is the crack length ($2a$=LOP) and t is the thickness of the plate.

3.11. Effect of Geometry in Load-Carrying Cruciform Joints

Table 3.2 lists different geometrical parameters were used including the attachment and main plate thickness, weld size, shape, and toe leg length (h, S). Moreover, the un-penetrated line LOP was treated as initial crack ($2a_i$). A uniform tensile stress range of 96 MPa was applied to the end of the attachment plate to consider the crack growth propagation from LOP.

Table 3.2: Geometrical parameters for load-carrying cruciform welded joints

B, (mm)	T, (mm)	$2a$, (mm) (LOP)	h, (mm)	S, (mm)	L_1, (mm)	L_2, (mm)
15	15, 25	3	6, 10	6, 10	150	150

3.11.1. Effect of Weld Shape

Due to the variation of throat thickness, the leg length either h or S, will be varied (see Figures 3.17 and 3.18). According to Örjasäter [93], the fatigue strength increases with the leg length, however, the influence of the leg length in this case is difficult to determine.

The weld shape usually depends on the welding position and welding conditions. Sometimes, the weld is shaped like a scalene triangle, or it has a concave or convex curvature. Hence, the applications of available solutions are limited.

The validity of Frank and Fisher equation has been compared with FE analysis in case of $h/S=1$ and $B/T=1$ where this equation can only be used as referred in Refs. [19, 84, 89]. Therefore, the present study aimed to investigate the effect of unequal weld toe length on main and attached plate on SIF. Moreover, the ability to use FRANC2D has been shown.

Figures 3.34 and 3.35 show the comparisons between SIF calculated by FRANC2D and those calculated from BSI (modified Frank, Fisher formula) for isosceles triangles shape fillet weld ($S=h$) and plate thickness ratio (B/T) of unity. It can be seen that the calculated results have very good agreement with BSI as discussed in the previous section. It shows that the increasing of weld size will decrease the SIF as shown in Figure 3.34 (a, and b).

In fact, the BSI solutions were estimated only for $h=S$, and $B=T$. Therefore, the current simulation showed that the empirical formulas are unable to find appropriate SIF solutions in case of unequal leg lengths ($h \neq S$), however, the ratio of h/B and S/B still being within (0.2-1.2), as shown in Figure 3.35. Hence, the FE solutions are considered.

It can be concluded that the leg length on the main plate side, h has a significant effect on SIF in the case of load-carrying cruciform joint more than that of the attached leg length, S. The SIF increases as leg length on main plate increases as shown in Figure 3.35. The same results of leg length effects have been obtained in case of $B \neq T$ as shown in Figures 3.36 and 3.37.

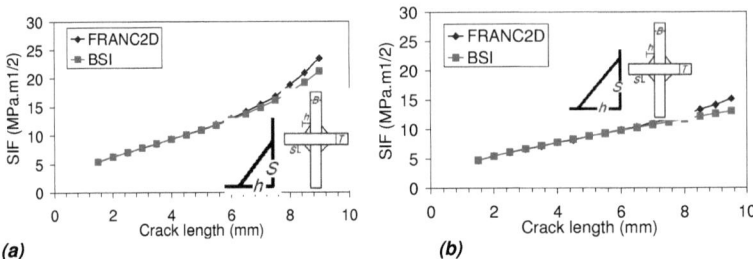

Figure 3.34: Calculated SIF for $B=T=15$ mm, $2a=3$ mm, isosceles triangles weld shape, $\Delta\sigma=96$ MPa: (a) $h=S=6$ mm; (b) $h=S=10$ mm

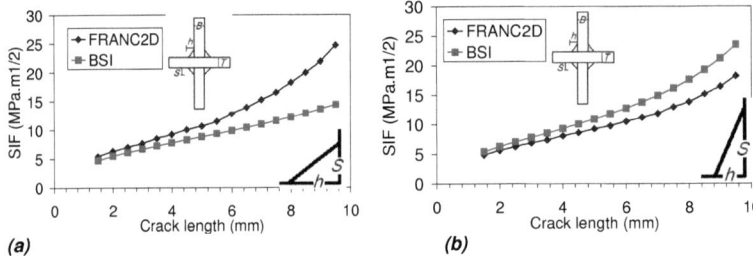

Figure 3.35: Calculated SIF for $B=T=15$ mm, $2a=3$ mm, non-isosceles triangles weld shape $\Delta\sigma=96$ MPa: (a) $h=10$ mm, $S=6$ mm; (b) $h=6$ mm, $S=10$ mm

3.11.2. Effect of Plate Thickness Ratio

The analytical formula from BSI and IIW are no longer able to determine the SIF solutions when $B \neq T$. The comparison with FEM for plate thickness ratio $B/T < 1$ are presented in Figure 3.36.

In isosceles triangles fillet weld shape there are no big differences between equal plate thicknesses $B/T =1$, (see Figure 3.34) and non-equal plate thicknesses $B/T < 1$ (see Figure 3.36). For design reasons, usually the ratio $B/T > 1$ is not considered [19].

It can be concluded that the effect of plate thickness on SIF and fatigue strength is less significant as compared with the effect of leg length. This is because the thickness affects different mechanisms in fillet welded joints in two types of cracks, the weld toe crack and the weld root crack. When fatigue cracks originate from a weld toe, the thickness affects the stress concentration at the weld toe. In contrast, only the thickness affects the initial crack size (LOP) when a crack originates from a weld root. Therefore, assuming of constant crack length (constant LOP), the effects of thicknesses are less. Unless the increasing in B will increase LOP, then the fatigue life will decrease.

No specific studies deal with the calculations and comparisons of SIF as based on FE, IIW and BSI to investigate the reliability of proposed solutions to use in fatigue life laws. Moreover, in knowledge of the author, FRANC was not applied to fillet welded joints cracking until now.

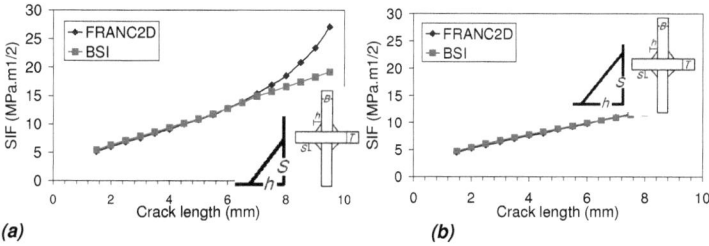

Figure 3.36: *Calculated SIF B=15, T=25 mm, 2a=3 mm, isosceles triangles weld shape $\Delta\sigma$=96 MPa: (a) h=S=6 mm; (b) h=S=10 mm*

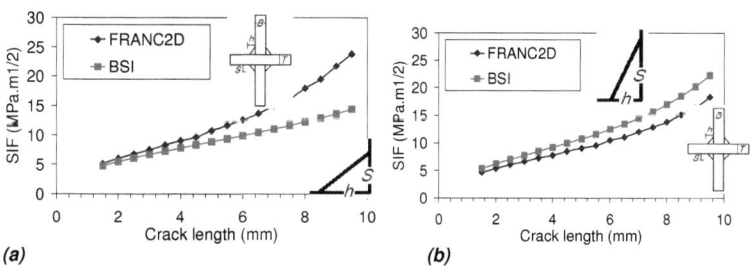

Figure 3.37: *Calculated SIF B=15, T=25 mm, 2a=3 mm, non-isosceles triangles weld shape $\Delta\sigma$=96 MPa: (a) h=10 mm, S=6 mm; b) h=6 mm, S=10 mm*

3.12. Conclusions

The main objective of this chapter is to show the reliability of FRANC2D program to calculate SIF in fillet welded joints. This factor will be used to calculate fatigue life. Several formulas were used to calculate SIF via crack length.

SIF at the weld toe and weld root of cruciform and butt welded joints have been evaluated using the FEM where the influence of the weld geometry was incorporated in the solutions. Meanwhile, the analytical SIF solutions have been derived for transverse butt joints having LOP.

The assumption was used that in the as-welded condition the crack remains open during the loading cycle due to the tensile residual stresses caused by welding are high enough. Therefore, the SIFs range corresponding to the nominal stress range is effective and independent of the R-ratio of nominal stresses. To include the effect of crack closure, other SIFs due to residual stresses should be calculated as will be discussed in Chapter Five. The FRANC2D software and quadrilateral elements were used to calculate the SIF of the joints from elementary of fracture mechanics.

This program has the ability to analyse a cracked body and to describe the singularity ahead of the crack tip. Thus, it can be concluded that for specific crack propagation, the SIF can be calculated under mode-I loading conditions. The efficiency of the current calculations was demonstrated for two examples of joints. In the present study, a developed analytical approach for toe crack in cruciform joints has been used. Frank and Fisher's formula was used in limited geometries in case LOP.

For the LOP forms of root cracks recommendations were given to use the modified solution from BSI 7910 instead of that from IIW. The reason for the difference between these two solutions (BSI and IIW) is that IIW uses the old Frank, Fisher formula whereas recently BSI modified this formula. These difference effects will be presented in Chapter 4 in fatigue life calculations.

The new modified solution from BSI is compared with that from FRANC2D and good agreements have been observed. However, FRANC2D has advantages over both solutions because it provides good crack path descriptions and considers the curved crack path. Hence, a more accurate SIF solution has been obtained. The crack path orientation angle has been investigated by using two crack propagation manners.

In the case of other geometries, the current solution of FE will be adequate to be used due to the limitation in available analytical and empirical solutions. The effect of weld toe length on main plate and cross plate was not sufficiently investigated and explained in literature. Therefore, this is another purpose of the present study which shows the effect of leg length on the SIF and crack propagation of fillet welded joints.

The main findings of the weld geometry effects are as follows:

1. The increase of weld size in isosceles triangles fillet weld shape will decrease the SIF.

2. The weld leg length, h and S have a major effect on SIF. The new ratio (S/h) was presented for SIF calculation in root crack of cruciform joints. The conclusions showed that the value of SIF is strongly affected by the leg length on the main plate side. The increasing h will significantly increase the SIF.

3. The effects of unequal plate thicknesses ($B/T < 1$) were studied, and there are no effects of plate thickness ratio on SIF solutions for load-carrying cruciform joints.

Chapter Four
FATIGUE LIFE CALCULATIONS AND VERIFICATIONS

4.1. Introduction

Fatigue life prediction of welded joints is very complex, costly and time consuming. This is due to its complex geometry, number of stress concentration points, and heterogeneous weld metal properties. To avoid the costly and complex procedure, fatigue life estimation of a joint for structural applications follows the S-N type of approach covered by recommendations and standards such as IIW [3, 34], BSI [14, 75, 76] GL [13], and Eurcode [12] etc. Though these recommendations have a considerable fatigue design data set, the studies of the effect of crack growth parameters C and m, initial crack length determinations and stress intensity factor calculations of welded structures are still not clear and have not been discussed enough.

To assess appropriately fatigue crack growth life of welded joints, it is necessary to obtain accurate results for the stress intensity factor (SIF) in the crack propagation phase.

In the previous chapter, the SIFs of weld notch cracks were calculated by FRANC2D program and the results have been verified. There is often a considerable amount of scatter in fatigue data even when carefully machined standard specimens out of the same lot of material are used. Therefore, a reduction factor is often applied to the S-N curves to provide conservative values of fatigue strength for the design of components that are called FAT class, listed in IIW, and are measured at two million cycles.

Moreover, the fatigue life, and the fatigue strength FAT of load-carrying and non-load-carrying welded joints were analyzed using linear elastic fracture mechanics LEFM.

Therefore, the aim of this work is to develop procedures to find the initial and final crack depth for welded geometries, and in turn to determine the FAT-values. In this work, FAT-values for some notch cases are calculated. Table 4.1 shows the geometries used in this Chapter. The geometry details also can be found in Table 3.1.

Thus, the new recommended limits of FAT for new geometries not listed yet in recommendations can be calculated according to backward calculations. Crack length parameters are determined. It will be shown that the calculated fatigue lives are in very good agreement with the existing experimental results.

Table 4.1: Welded joints used in this chapter

Transverse butt weld	
	The weld edges machined or ground flush to plate. Double side complete weld penetration with toe crack.
	Weld reinforcement with double side complete penetration with toe crack.
	Lack of penetration (LOP) being considered as a root crack.
Cruciform joint or T-joint	

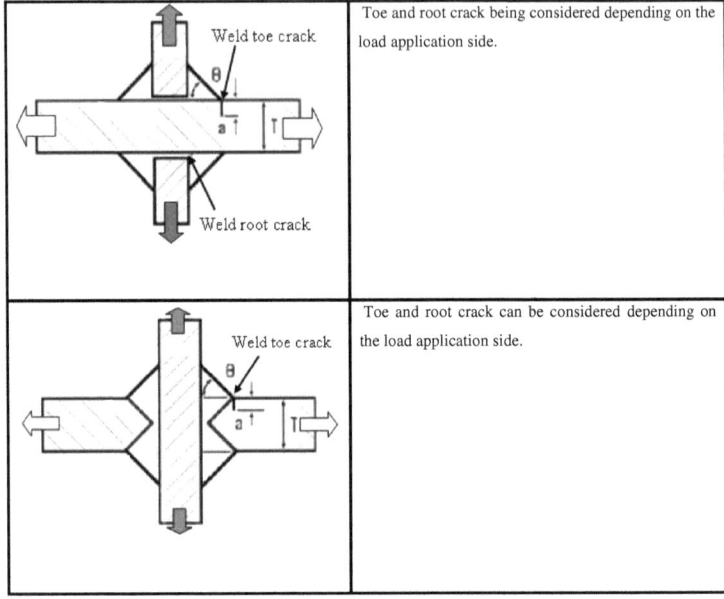

4.2. Fracture Mechanics Analysis

At present, there are two primary approaches used for predicting fatigue life namely, the fracture mechanics approach and the S–N curve approach. Figure 4.1 describes the relationship between these two approaches [23]. Fracture mechanics is mostly used in life prediction of a structure with an existing crack, and is based only upon crack-growth data.

The initiation phase is assumed negligible for welded joints in the fracture mechanics approach and the life is based upon a SIF, which accounts of the magnitude of stress at crack tip, crack size and joint geometries.

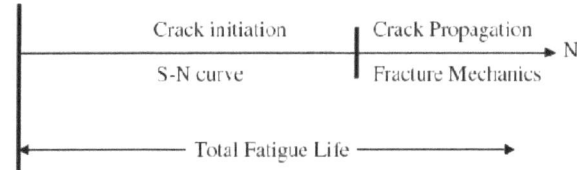

Figure 4.1: *Relationship between the characteristic S–N curve and fracture mechanics approaches [23]*

The stress value used in some guideline is the so-called nominal stress level which is a stress value remote from the weld. This stress is then connected to the weld class system using a factor for the actual type of weld and quality level. In principal this is a good system, but there are many practical difficulties involved in applying the system due to the difficulties and impossibilities to define in complex structures, where many points for stress concentrations and different loading conditions are present [32, 36].

For complex welded structures with many attachments and loads, fracture mechanics based methods can be used to determine fatigue strength of an unknown notch case of welded connections. The actual weld toe geometry is considered and an initial crack is assumed. Moreover, the *S-N* model cannot deal with the presence of cracks; therefore, fracture mechanics is an indispensable tool in situations where a crack is detected and sized [1].

Finally, to be able to use the nominal stress method, the current structure must be similar to one of the structures available for the method. Misalignments and defects must lie within the weld class, which means the nominal stress method is only applicable to the toe side of the weld, and the root side of the weld is frequently neglected in the design process [36]. In case of welded root cracking, it is usually not possible to measure the stress in the vicinity of the initiation point since the crack initiation location is embedded inside the weld [94].

All the above mentioned drawbacks using the nominal stress method in design can be avoided if local methods are being used. There are at least two such methods: the so called effective notch method and linear fracture mechanics, which both have proven to work well. In these methods, the results are taken in direct connection to the weld and all drawbacks in the nominal method are avoided [32, 36].

The above-mentioned local methods are typically more accurate than the nominal stress method, especially as the geometric complexity of a structure increases.

4.3. Fatigue Life Calculations Using Paris' Law

Several equations are proposed by different researchers, usually semi- or wholly-empirical, to correlate fatigue crack growth rate data with the range of the single parameter ΔK, the range of SIF. Among the proposed equations, Paris–Erdogan relationship is commonly accepted and used in practice for a wide range of mode-I crack [77]. This equation is also recommended by IIW to calculate fatigue life for steel and aluminium welded joints and is still in use today for many applications:

$$\frac{da}{dN} = C(\Delta K_I)^m \tag{4.1}$$

where da/dN is the crack growth rate per cycle, C and m are material constants determined from a curve fit (da/dN-ΔK curve) to the test data, and ΔK_I is the range of the SIF opening mode [15]. The SIF will be calculated from a FEA as described in the previous chapter. Paris' law can be used when it is known that the welding process introduces inherent surface crack-like flaws at the weld toe, i.e., along the fusion lines when these lines are brittle and at the weld root. These flaws are regarded as initiated cracks, which is a somewhat conservative assumption [31].

The calculated fatigue lives are compared with the real experimental test results. In addition, a fracture mechanics analysis based on accepted standards is done as a reference.

With this local stress fracture mechanics approach, Paris' equation can be used to appropriately predict the fatigue crack propagation N_p and fatigue life N_T of a welded structure.

The total life N_T of a welded joint consists of two phases, namely the crack initiation life N_I and the crack propagation life N_P, where the initial crack grows to a certain point and affects the safety of the structure.

Furthermore, crack initiation in fracture mechanics is defined as a number of cycles that cause a certain length of crack. Hence, because of welded joints may contain inherent crack-like defects, discontinuities and stress concentration points, the crack initiation life is assumed to be negligible.

In this work the determination of the number of cycles N until failure is done by integration for the crack growth relation from the initial crack depth (a_i) up to reaching a critical crack depth at break-through (a_f) with the use of the Paris-Erdogan equation:

$$dN = \frac{da}{C(\Delta K_I)^m} \Rightarrow \int_0^N dN = \int_{a_i}^{a_f} \frac{da}{C(\Delta K_I)^m} \qquad (4.2)$$

The SIF range inside is:

$$\Delta K_I = Y \Delta \sigma \sqrt{\pi a} \qquad (4.3)$$

where $\Delta \sigma$ is the applied stress range, a is the crack length, and Y is the correction factor mentioned in Chapter Three as a function of $f(a/t)$.

$$\int_0^N dN = \frac{1}{C} \int_{a_i}^{a_f} \frac{da}{\left[Y \Delta \sigma (\pi a)^{\frac{1}{2}} \right]^m} \qquad (4.4)$$

Thus, the number of cycles for one increment is:

$$N = \frac{1}{CY^m (\Delta \sigma)^m \pi^{\frac{m}{2}}} \frac{\left[a_{i+1}^{(1-\frac{m}{2})} - a_i^{(1-\frac{m}{2})} \right]}{1 - \frac{m}{2}} \qquad (4.5)$$

In the above equations, the geometry factor Y is assumed as constant, since the inclusion of a function of (a/t) within the integral sign will usually lead to a formulation that cannot be integrated analytically. Moreover, the integration cannot be performed directly since the geometrical factor changes with a. Therefore, to calculate the geometry factor Y in the above formulas, the SIF has to be evaluated. This is done numerically by using the FEM. Then, according to SIF, the number of cycles N can be calculated for different loads.

Paris' equation (Eq. 4.1), which is used to calculate fatigue life in FRANC2D, is very simple and may not be appropriate for some materials, nonzero load ratios, and very high or very low SIF ranges. Moreover, the simple application of Paris' equation does not take into account the effects of residual stresses and crack closure, where the latter effects are being included within the stress ratio R. Thus, in many cases, it is more appropriate to calculate the extract SIF vs. crack length history within FRANC2D, and use this information with a more sophisticated growth model [78, 79].

Then, other growth equations can be used as Forman, Walker and NASGRO. Finally, the addition of the crack closure phenomenon to Paris' model is successfully solved.

In practice, Paris' law is more straightforward and very often sufficiently accurate to solve the fatigue life equation by splitting the crack growth history into a series of crack increments (Δa) in which Δa is determined according to the stability of the a-N curve.

Through dividing the crack path into (n) increments, between the initial size a_i and the final size a_f, then the number of cycles N_j for each increment (between the size a_j and the size $a_j+\Delta a$) can be calculated. The crack length vector and calculated SIF are transferred to Excel and integrated numerically according to Eq. (4.2). Then, the total life N can be calculated by summation of the N_j for each increment as:

$$N = \sum_{j=1}^{n} N_j = \sum_{j=1}^{n} \frac{\Delta a}{C(\Delta K_I)_j^m} \tag{4.6}$$

where j is the steps number. Thus, a numerical integration for a crack growth rate is carried out at various stress levels, and the results of fatigue life are recorded to determine the *S-N* curve. This original model (Model-A) is still used today for many applications.

More advanced models essentially build on the Paris' equation by addressing mean stress effects, threshold behavior (Region-I) of da/dN-ΔK curve, the instability asymptote (Region-III), and fatigue crack closure effects. The NASGRO equation is a full range model that mathematically includes all three regions of FCG curve (Model-B).

4.4. Fatigue Life Calculation Using NASGRO Equation

The superposition method originally is based on the principle of LEFM [65]. Under the cyclic load and residual stresses, the SIF range and stress ratio are calculated as:

$$\Delta K = (K_{max,app} + K_{res}) - (K_{min,app} + K_{res}) = K_{max} - K_{min} = \Delta K_{app} \quad (4.7)$$

$$R_{res} = (K_{min,app} + K_{res})/(K_{max,app} + K_{res}) \quad (4.8)$$

Therefore, under cyclic loads, only R_{res} changes due to the presence of residual stresses. Considering the welding residual stress effect the original NASGRO equation is expressed as:

$$\frac{da}{dN} = C_{ESA}\left[\left(\frac{1-f}{1-R}\right)\Delta K\right]^n \frac{\left(1-\frac{\Delta K_{th}}{\Delta K}\right)^p}{\left(1-\frac{K_{max}}{K_c}\right)^q} \quad (4.9)$$

where f is Newmans effective stress ratio and ΔK_{th} is the threshold value of ΔK for a given applied stress ratio $R=K_{min}/K_{max}$. The parameter C_{ESA}, and the exponents n, p, q determine the shape of the da/dN curve. K_{max} is the stress intensity factor at the maximum load for a given R ratio, and K_c is the critical stress intensity factor which relates to the thickness of the specimens; more details can be found elsewhere [95].

The solutions, then, use implemented SIF solutions (FEM) and crack growth concepts to propagate the crack until failure occurs. The Newman's equation for f is defined as:

$$f = A_{0,n} + A_{1,n}R + A_{2,n}R^2 + A_{3,n}R^3 \tag{4.10a}$$

For $R \geq 0$, and for $-1 \leq R < 0$ as;

$$f = A_{0,n} + A_{1,n}R \tag{4.10b}$$

The polynomial coefficients are defined as:

$$A_{0,n} = (0.825 - 0.34\alpha + 0.05\alpha^2)\left[\cos\left(\frac{\pi}{2}S_{max}/\sigma_0\right)\right]^{1/\alpha} \tag{4.11a}$$

$$A_{1,n} = (0.415 - 0.071\alpha)S_{max}/\sigma_0 \tag{4.11b}$$

$$A_{2,n} = 1 - A_{0,n} - A_{1,n} - A_{3,n} \tag{4.11c}$$

$$A_{3,n} = 2A_{0,n} + A_{1,n} - 1 \tag{4.11d}$$

where α is a material constant, its value varies form 1 for plane stress to 3 for plane stain; S_{max} is the peak stress of a stress cycle, and σ_0 is the material flow stress (i.e., the average of the tensile yield and ultimate strengths) [96].

NASGRO equation incorporates fatigue crack closure analysis for calculating the effect of the stress ratio on the crack growth rate under constant amplitude loading.

The threshold stress intensity factor range in Eq. (4.9), ΔK_{th}, is approximated as a function of the stress ratio, R, the Newman's closure function f, the threshold stress intensity factor range at $R = 0$, ΔK_0, the crack length, a, and an intrinsic crack length, a_0, by the following empirical equation:

$$\Delta K_{th} = \Delta K_0 \left(\frac{a}{a+a_0}\right)^{1/2} / \left(\frac{1-f}{(1-A_0)(1-R)}\right)^{(1+C_{th}R)} \quad (4.12)$$

The values of C_{th} and ΔK_0 are stored as constants in the NASGRO materials files, and a_0 has been assigned a fixed value of 0.0015 in. (0.0381 mm). The materials data necessary for the used materials are obtained from AFGROW database [97] and Ref. [98] (see Table 5.1 of Chapter Five). The SIF due to maximum applied load is defined as:

$$K_{max} = \frac{\Delta K}{(1-R)} \quad (4.13)$$

where ΔK varies between ΔK_{th} and K_{max}.

The current calculation using Model-B becomes more complex for structures which have the distributions of residual stresses. The use of NASGRO equation under the effect of residual stresses (Model-C) will be discussed in the next chapter.

4.5. Parameters of Crack Propagation Life

The problem that arises to determine the fatigue life is to choose the appropriate parameters of C, m, a_i, and a_f. Some studies have presented a range of length for these crack-like defects in welded joints, and they have given conservative values of lives. Due to the complex joint geometry, a number of stress concentration points and heterogeneities of weld metal properties exist in addition to the effect of residual stresses.

The accurate crack lengths and growth parameters have not been established yet. And this is one of the purposes for the present work.

Traditionally, the fatigue design of welded joints for structural applications has used the S-N curve type of approach based on experimental results for different weld geometries [54], included for example in Eurocode 3, BSI7910, BS5400, BS7608 [12, 14, 75, 76] and IIW [3, 34] where the initial crack length is non-measurable yet and no guidance is found. However, initial cracks used in fatigue analyses are often in the range of 0.05–0.2 mm [29]. Nevertheless, these values can vary depending on the welding operation parameters, geometry and materials properties.

Engesvik [30] has also analyzed the fatigue life of welded joints and concluded that it may be dubious to apply LEFM at crack depths less than 0.1 mm. In other literature, a_i is usually measured or approximated to 0.1-0.2 mm for welds [31].

BS7910 [14] recommended the initial flaw size, a_i between 0.1 and 0.25 mm. The life is assumed to be finished when the final crack, a_f reaches one-half of the sheet thickness [31, 32]. Therefore, in this work, trails are carried out to clamp down on bogus sizes of a_i and a_f. The parameters a_i, and a_f are determined when the C, and m were fixed. Moreover, the actual weld toe geometry is considered and an initial crack is calculated.

4.6. Selection of the Notch Cases

Cruciform and butt welded joints are the most used geometries in engineering structures. The factors that presently inhibit the full use of cruciform components are the lack of standard and suitable specimen design. This aspect is now being addressed through FEM program with help of fracture mechanics approach through this work to find the values of the fatigue life and initial crack depths for cruciform weld geometries. The new recommended limits of FAT and initial crack growth parameter are calculated.

There are two locations of crack initiation that exist at cruciform joints with fillet welds, at the weld toe, transition between weld and plate, and at the weld root, see Figure 4.2. The first crack will propagate through the base plate, whereas the second one will propagate through the weld throat [3, 33].

Chapter Four *Fatigue Life Calculations and Verifications*

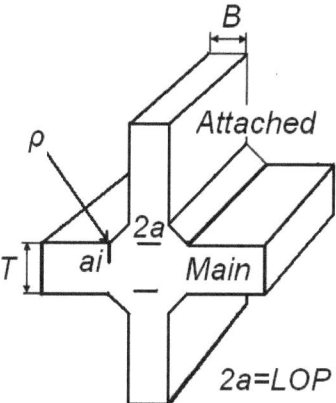

Figure 4.2: Cruciform fillet welded joint

The cruciform weld joint geometry with ratio of attachment plate to main plate thickness $B/T=1$, is shown in Figure 4.2, where ρ is weld toe radius, a_i and $2a$ are the initial crack length located at weld toe and LOP, respectively. B is the attached plate thickness and T is the main plate thickness. The advantage is taken of symmetry to model quarter of the complete geometry. Two specimens are modelled; non-load-carrying cruciform where load acted at the end of main plate and load-carrying cruciform joint where the load acted at the end of attached plate (see Figure 4.2).

In the same manner, two butt welded joints are calculated; first butt joint with toe crack and second butt joint with LOP. In latter LOP is considered as a root crack. By contrast, in former toe crack is considered to start from the weld toe.

4.7. Backward Calculation of Fatigue Strength

In this chapter, fatigue life calculation procedures are carried out using a simple form of Paris' law as based on fracture mechanics method. With knowledge of the SIF, K_I at different crack depths, it is possible to make polynomial curve fits of $K_I(a)$. Numerical integration for Paris' law (Eq. 4.1) is carried out using a polynomial equation to establish the *a-N* curve. SIFs at different loading have been scaled to calculate the expected fatigue life for different stress ranges and to develop the *S-N* curve. A FORTRAN language program is developed to carry out the numerical integration of Paris' law at various stress ranges. The initial crack length is changed to get characteristic and mean FAT values. In this work, these calculations are referred to Model-A.

If the correct constants for steel (DOMEX 550 MC, high-strength hot rolled steel with the minimum yield strength 550 MPa and the tensile strength minimum 600 MPa and maximum 760 MPa, respectively) were not available, the BSI7910 [14] and IIW [3, 34] recommendations were used, as shown in Table 4.2.

Table 4.2: Characteristic values of recommended materials parameters for steel. Units used are in N and mm

BSI 7910	IIW
$C_{95\%}=5.21\times10^{-13}$, and $m=3$	$C_{95\%}=5\times10^{-13}$, and $m=3$

Most of fatigue results are distributed and scattered around the mean life value and further away. If these data are plotted, it will have a bell-shaped curved. The standard deviation (*STDV*) is given for data that are distributed normally. According to IIW, all fatigue resistance data (FAT values) are given as characteristic values, which are assumed to have a survival probability (reliability) of at least 95% (i.e., 5% failure probability) within two standard deviations equal to 0.178 calculated from the mean value of a two-sided 75% confidence level [3]. The initial crack length should be considered in determination of fatigue life of welded joints. Emphasis is laid on how to choose initial crack length, a_i with the fixed crack growth parameters (*C*, and *m*). With the backward calculations the initial crack length value has been determined which coalescence with the FAT95% according to characteristic value of C and m. In case of FAT50%, a new value of $C50\%$ is needed to be determined, which is equal to $C95\% \pm 2STDV$. Both curves of these FAT values are plotted, as shown in Figure 4.3, using the straight line equation with slop m, i.e.,:

$$LogN = LogC - mLogFAT \qquad (4.14)$$

Then the characteristic fatigue strength FAT95% is calculated by:

FAT95%: $Log(2 \times 10^6) = LogC\,95\% - mLogFAT\,95\%$ \qquad (4.14.1)

The mean fatigue strength FAT50% is then:

FAT50%: $Log(2 \times 10^6) = Log(C\,95\% + 2 \times STDV) - mLogFAT\,50\%$ \qquad (4.14.2)

Finally, the *S-N* curves are obtained (see Figure 4.3) according to:

$$N = (FAT\,95\% / \Delta\sigma)^m \times (2 \times 10^6) \qquad (4.15)$$

$$N = (FAT\,50\% / \Delta\sigma)^m \times (2 \times 10^6) \qquad (4.16)$$

The mean values of materials constants calculated in this work are given in Table 4.3 for steel (DOMEX 550 MC steel [31]). The value of $C50\%$ is only 8.5% larger than that from BSI 7910 which is given equal to 2E–13 [14]. These values can be used to calculate FAT50% for new notch cases.

Table 4.3: Mean values of recommended materials parameters for steel. Units used are in N and mm

BSI 7910	Current work
$C_{50\%}=2\times10^{-13}$, and $m=3$	$C_{50\%}=2.17\times10^{-13}$, and $m=3$

| Chapter Four | Fatigue Life Calculations and Verifications |

According to the current procedures using Model-A (backward calculations), the crack length value is changed just to fit the curve of FAT95% which is plotted using Eq. (4.15), as shown in Figure 4.3. Then, these determined values (C, m) from Table 4.1 and initial crack length from the backward calculations can be used to find FAT95% for new notch cases which have the same crack classifications (weld toe or weld root crack). Curve of FAT50% is plotted using Eq. (4.16) and values of $C50\%$ and m equal to 2.17E–13 and 3, respectively (see Table 4.3).

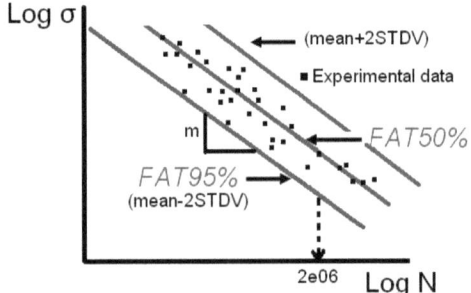

Figure 4.3: S-N curve

4.8. Results and Discussion

SIFs and fatigue life have been calculated for welded joints using fracture mechanics. The joints are assumed to be free from residual stresses, undercuts and other weld defects. The fatigue test results for joints which failed from weld toe and LOP are taken from the works of Lindqvist [31] and Singh et al. [84], respectively. The effect of residual stresses will be discussed within the next chapter.

4.8.1. Fatigue life Calculations

In the following, the *S-N* curves of characteristic and mean fatigue life, FAT95% and FAT50%, respectively, are shown. FAT50% refers to the experimental mean fatigue strength value that is applied during the real service time for steel structures using the calculated $C50\%$ equal to 2.17E–13 and $m=3$ (i.e., 50% failure probability, 50% survival probability). The current numerical integration refers to the backward calculation using Paris' law.

The initial crack has been determined which gives the credible coalescence with FAT 95% class from the recommendations. The value of a_i is 0.1 mm for the crack initiated from the weld toe, while a_i is equal to the un-penetrated line for the joints having LOP or incomplete melting.

To verify the current calculations (Model-A), the results are then compared with experimental data. It is found that these initial crack lengths of each case are uniform and have been confirmed for all types of joints. In contrast, final cracks defined in many researches to be equal to one-half of the parts thickness. In case of LOP, the a_f was set to be 0.8× (leg length on attached plate side) + (attached plate thickness)/2. The coefficient multiplying leg length is varied between 0.6 and 0.9 [49]. In all cases, a_f has a less significant effect on the fatigue life [31, 40], and the variation can be considered as negligible.

4.8.1.1. Cruciform Weld Joints having Toe Crack

To determine fatigue life curves for the particular weld details, various standard groups are put together into categories with similar fatigue behaviour. The compiled results are from fatigue life testes in several large experimental investigations carried out in Europe, where the applied stress ranges are in the region of 80 to 105 MPa and the thickness of the plates is from 16 to 38 mm and stress ratio R between 0 and 0.3 [1]. Linqvist [31] tested experimentally 12 mm non-load-carrying cruciform specimens under tension made of the steel DOMEX 550 MC, which is an extra high-strength hot rolled steel with minimum yield strength 550 MPa. The tensile strength is minimum 600 MPa and maximum 760 MPa, respectively. The weld toe radius is set at 0.6 mm. The model started with the initial crack a_i from the weld toe. The initial crack will grow in depth regularly. The reported test results are compared with calculated results from backward calculations procedure used in the current study.

Fatigue life calculation is defined as the stage when the crack has reached its final length, which is defined in many researches to be equal to one-half of the parts thickness [31]. In this work, a_f values equal to one-half of the sheet thickness is adopted. In addition, these values of a_f have been verified which give the fix number of cycles in the a-N curve, i.e when the increase in fatigue life was negligible [15]. Figure 4.4 shows that a_f is about 5.2 mm in case of sharp weld toe (no radius). The value of a_f increases to approximately 6 mm when the weld toe radius increases to 0.6 mm. In contrast to the influence of a_i, the fatigue life is not as sensitive to the values of the a_f.

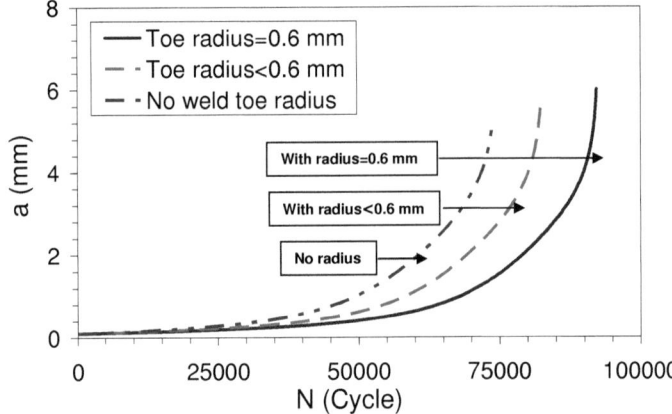

Figure 4.4: *Crack length as a function of number of cycles calculated using Paris' model*

The possibilities to study the effect of local geometry have been demonstrated (see Figure 4.4), where the effect of toe radius on fatigue life calculation is shown. Decreasing the weld toe radius makes the stress concentration more obvious, and thus decreases the fatigue life. In the recommendations, the value of fatigue strength FAT did not include the effect of weld toe radius. Hence, the current calculation (Model-A) presents an advantage to calculate the effect of different weld geometries, which have not been listed yet in the standards and recommendations.

The *S-N* curve, Figure 4.5, shows the backward calculations as compared with IIW for class 413 [3] for an incompletely penetrated load-carrying cruciform joint. The FAT95% class 413 in absence of toe radius is close to the currently calculated values as based on fracture mechanics. The initial crack length is calculated and identified for this case equal to 0.1 mm, because this value gives good results compared with FAT95% (63 MPa).

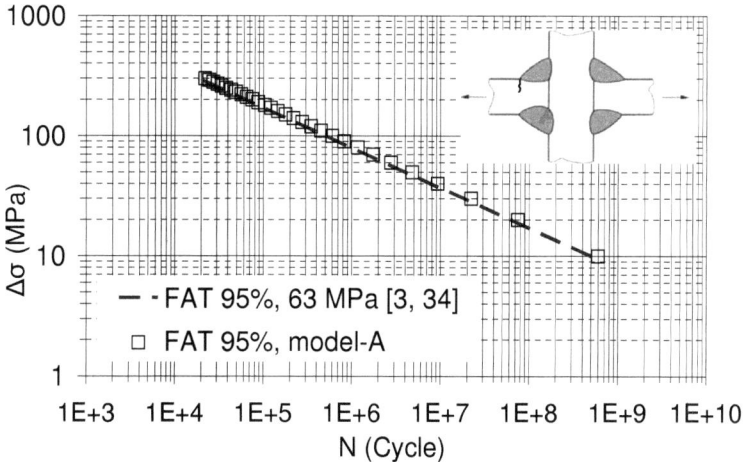

Figure 4.5: *S-N curves for incompletely penetrations cruciform joint having toe crack (a_i=0.1 mm, C_{char}–5E-13, m=3) compared with case 413, FAT63 from IIW [3, 34]*

The results of a fully penetrated load-carrying cruciform joint failing from the weld toe (FAT 95%, 71 MPa, class 412) are presented in Figure 4.6.

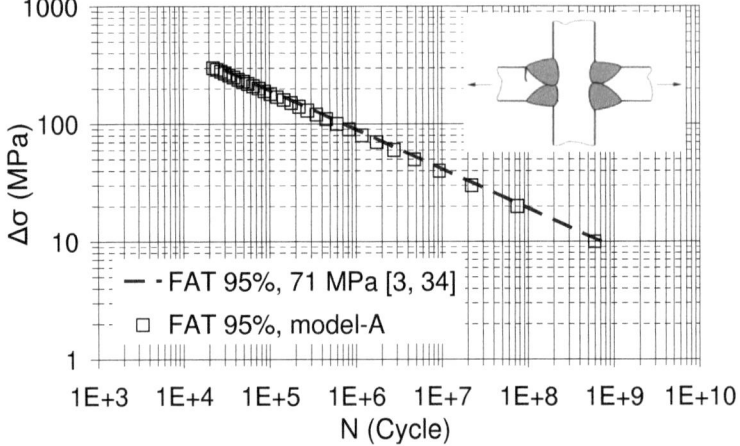

Figure 4.6: *S-N curves for full penetration cruciform joint having toe crack ($a_i=0.1$ mm, $C_{char}=5E-13$, $m=3$) compared with case 412, FAT71 from IIW [3, 34]*

From Figures 4.5 and 4.6, it is noticed that the value of initial crack 0.1 mm is creditable for the case of toe crack in two types of cruciform joints. The currently calculated life is located between these two lives. The calculated lives, characteristic and mean life (FAT95% and FAT50%, respectively) are presented in Figure 4.7, in comparison with the values from IIW.

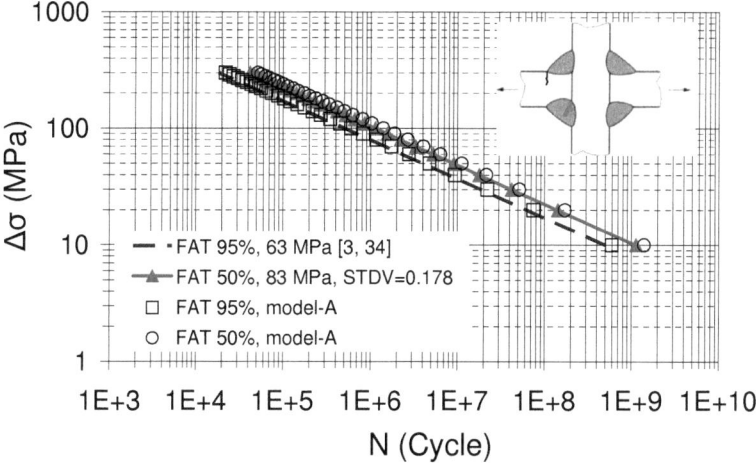

Figure 4.7: S-N curves for incomplete penetrations cruciform joint having toe crack (a_i=0.1 mm, C_{char}=5E-13, C_{mean}= 2.17E-13, m=3) compared with case 413, FAT63 from IIW [3, 34]

4.8.1.2. Transverse Butt Weld Joints having Toe Crack

The toe crack case is verified for transverse butt weld joints with toe crack as shown in Figure 4.8. The initial crack, a_i is equal to 0.1 mm. The complete penetration has been assumed and only toe crack is considered.

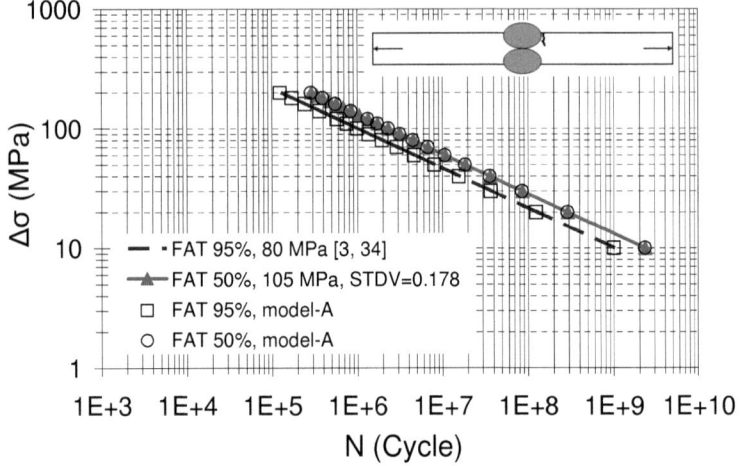

Figure 4.8: S-N curves for full penetration transverse butt weld having toe crack ($a_i=0.1$ mm, $C_{char}=5E-13$, $C_{mean}= 2.17E-13$, $m=3$) compared with case 213, FAT80 from IIW [3, 34]

4.8.1.3. Cruciform Weld Joints having LOP

The other type of crack is formed during the fabrication where a non fused gap will initiate the crack called lack of penetration LOP. The LOP size has a strong effect on the fatigue life of load-carrying cruciform joints.

Germanischer Lloyd Aktiengesellschaft GL [13] presented some recommended values of fatigue strength for welded metal in load-carrying fillet welds at cruciform or tee joint in case of LOP. These values from GL are based on the stress range in weld throat, and differ from that in IIW. FAT values for steel are 36 MPa for throat thickness > (plate thickness/3), and 40 MPa for throat thickness < (plate thickness/3). Regardless of the weld size, IIW stated that FAT for steel is 45 MPa [3, 34].

The FAT values from GL in case of load-carrying cruciform joint have provided more realistic results, as compared with calculated values in this study. Figures 4.9 and

Chapter Four   Fatigue Life Calculations and Verifications

4.10 show the comparisons of these FAT values with the current approach (Model-A). The fatigue strength increases with a decrease of weld size (throat thickness) due to the effect of the stress concentration.

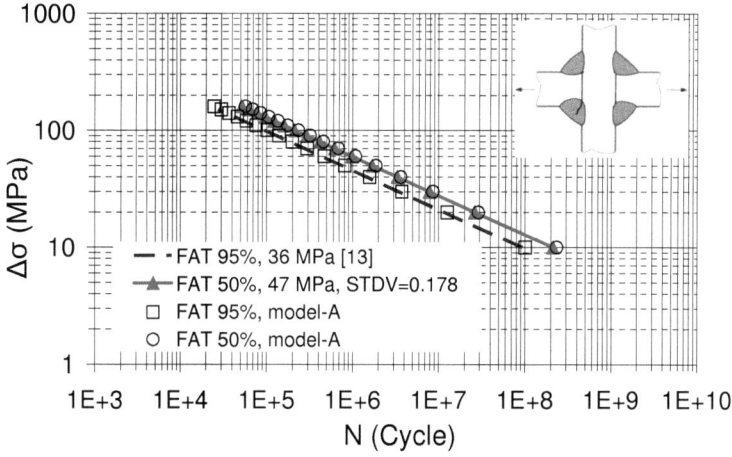

Figure 4.9: S-N curves for incomplete penetration cruciform joint having LOP ($2a_i$=LOP=7 mm, C_{char}=5E-13, C_{mean}= 2.17E-13, m=3 and throat thick>t/3) compared with type No. 23, FAT36 from GL [13]

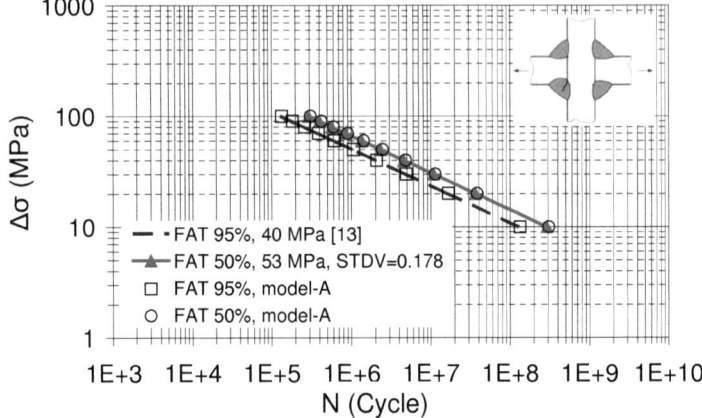

Figure 4.10: *S-N curves for incomplete penetration cruciform joint having LOP ($2a_i$=LOP=4 mm, C_{char}=5E-13, C_{mean}= 2.17E-13, m=3 and throat thick<t/3) compared with type No. 23, FAT40 from GL [13]*

Recall to the discussion in section 3.10.4.2 of SIF calculation of load-carrying cruciform joints where the difference between FRANC2D, BSI and IIW was presented (see Figure 3.31). Figure 4.11 shows the fatigue life curves for these different SIF solutions. IIW shows a higher FAT solution in this case as compared with BSI and FRANC2D (fracture mechanics).

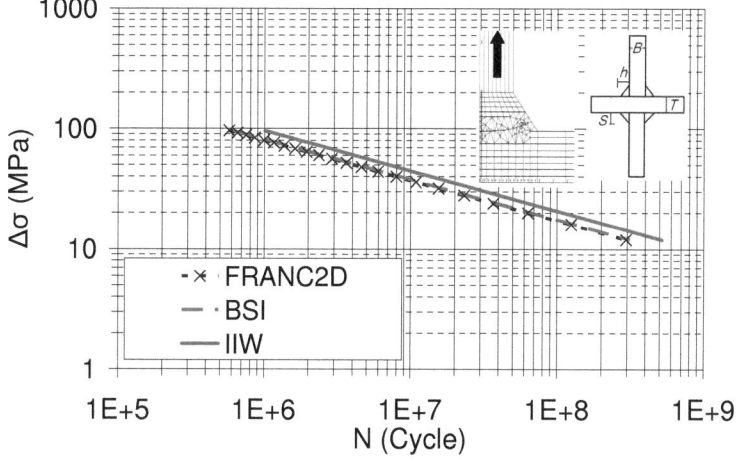

Figure 4.11: S-N curves from different SIF solutions in incomplete penetration cruciform joint having LOP, $C_{char}=5E-13$, $m=3$), $2ai=LOP=3$ mm, $B=T=15$ mm, $h=S=6$ mm

4.8.1.4. Transverse Butt Weld Joints having LOP

For butt weld joints welded from both sides, it is difficult to ensure that full root penetration is achieved. Fatigue failure from the weld roots may result in unexpectedly low fatigue strength [15]. Therefore for the same butt weld joint, FAT for LOP is lower than that for toe crack (45 and 80 MPa, respectively). Keep in mind that the latter FAT80 considers the full weld penetration (no LOP).

The *S-N* curves of butt weld joints with LOP are presented in Figure 4.12. An agreement is obtained when a_i equals to the LOP defect which existed in the beginning. Weld overfills are not taken into consideration.

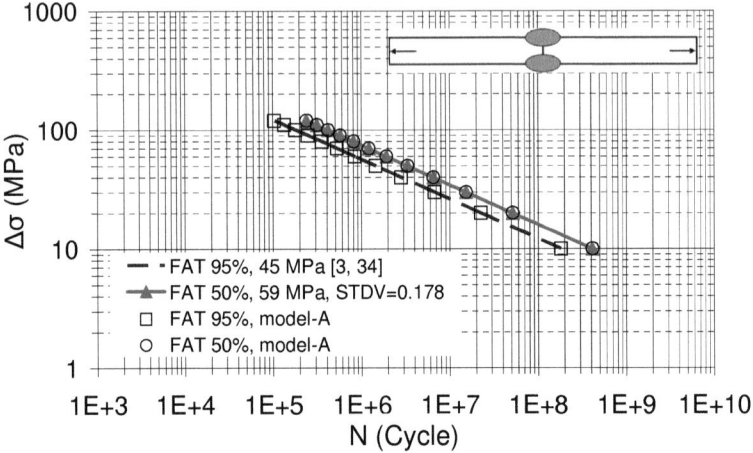

Figure 4.12: S-N curve for transverse partial penetration butt weld joint having LOP (LOP=2a$_i$=2.2 mm, t=10 mm, LOP/t=0.22 C$_{char}$=5E-13, C$_{mean}$= 2.17E-13, m=3) compared with case 217, FAT45 from IIW [3, 34]

Nykänen et al. [15] predicted FAT classes for common welded joints. The predicted fatigue strength of transverse butt weld is calculated using Eq. (4.17) as follows [15]:

$$FAT_{DOB} = \sum_{i=1}^{35} A_{DOB,i} \left(\frac{LOP}{t}\right)^{a_i} \left(\frac{H}{W}\right)^{b_i} \left(\frac{l_1}{l_2}\right)^{c_i} \left(\frac{W}{2l}\right)^{d_i} f(t) \quad (4.17)$$

Dimension variables of the butt weld are shown in Figure 4.13. The variables are the degree of weld penetration, LOP, weld heights l_1, l_2, weld widths W_1, W_2, and weld leg heights $H1$, $H2$. The weld shape was chosen so, that $W_1/2l = W_2/2l = W/2l$ and $H_1/W_1 = H_2/W_2 = H/W$.

The coefficients $A_{DOB,i}$ corresponds to the DOB being evaluated. a_i, b_i, c_i, d_i are exponents as listed in Ref. [15].

To make a valid comparison between current the calculation (Figure 4.12, where t=10 mm, LOP=2a$_i$=2.2 mm, and LOP/t=0.22) and Näkenen [15] (where t=25 mm), the constant ratio LOP/t=0.22 was used to be constant between them. Figure 4.14 shows the comparison between the current work and Eq. (4.17) for the two different thicknesses 10 mm and 25 mm as used in the current work and Nykänen's work, respectively. The calculated FAT from the current work using Model-A shows good correlation with predicted fatigue strength (FAT) from Nykänen (Eq. 4.17).

These results are approved with the current approach using the calculated crack length and SIF from this work.

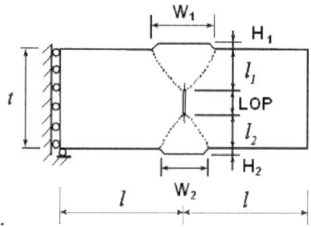

Figure 4.13: Transverse partial penetration butt weld with root cracks only [15]

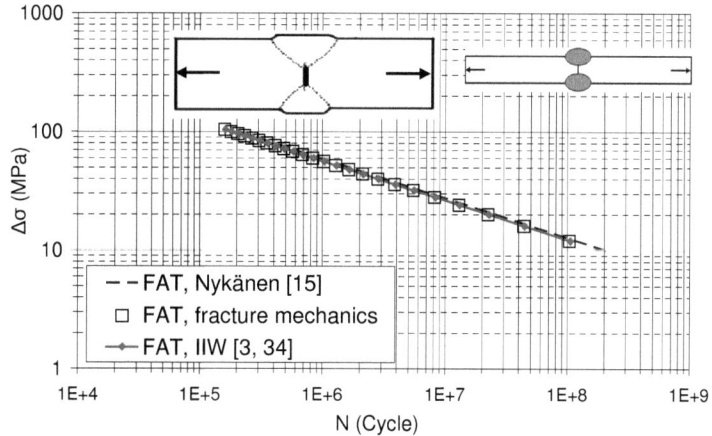

Figure 4.14: S-N curves comparisons in case of LOP of transverse butt weld, LOP/t=0.22

Chapter Four **Fatigue Life Calculations and Verifications**

The FE direct simulation for butt weld joint from Nykänen (Figure 4.13) is carried out using FRANC2D where the thickness=25 mm. Then, SIF was calculated and Eq. (4.18) is developed to be used in Paris' numerical integration program.

$$K = 0.2914(a)^4 - 5.8122(a)^3 + 45.708(a)^2 - 100.86(a) + 344.57 \qquad (4.18)$$

The initial crack equals to the LOP is assumed using Model-A, with Paris' law constants $m = 3$, and $C_{char} = 3$ E-13 (da/dN in mm/cycle and ΔK in Nmm$^{-3/2}$) as used by Nykänen [15], the fatigue life curves are presented in Figure 4.15. Good agreement between the two solutions is shown in this case.

It is to be emphasized that the predicted FAT from Eq. (4.17), are calculated for $H/W = 0.1$, $W/2l = 0.5$ and $t = 25$ mm. This mean that for the unequal weld dimension ($W_1 \neq W_2$, $l_1 \neq l_2$), Eq. 4.17 is not able to find a corresponding FAT. Therefore the recommendation to use the current approach (Model-A) with the calculated parameters will take an advantage, because only SIFs with a_i are needed to find an appropriate FAT for different geometries (see Figure 4.15).

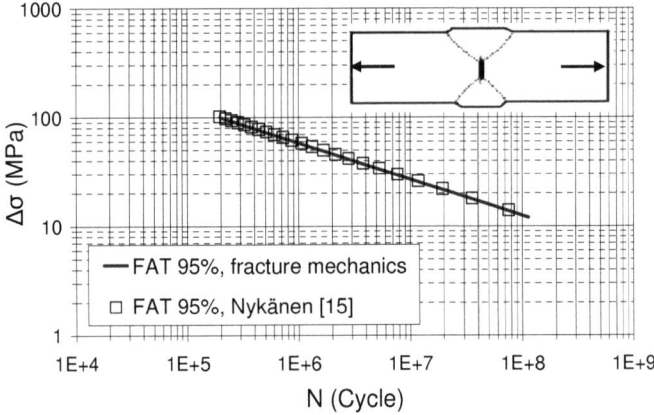

Figure 4.15: *S-N curves using Eq 4.17 and the current approach (Model-A) for LOP of transverse butt weld, thickness=25 mm, $2a_i=LOP=5.5$ mm, $LOP/t=0.22$*

4.9. Experimental Verifications

From literature [16, 17, 18], it is evident that most of fatigue life predictions of fillet welded joints are based on toe failure. Some other studies [20, 22, 84, 92] have considered the fatigue behaviour of fillet welded cruciform joints failing from the root region.

For fillet welds, the high stress concentration locates at the weld toe due to the fact that these locations are rarely sound and usually weldment contains flaws, crack-like defects and stress concentrations. Therefore, the presence of the weld toe radius inevitably will reduce these concentrations of stresses near the weld toe.

The effect of the toe radius is not included in FAT values from IIW and BSI 7910 recommendations. This will provide a good example to evaluate the current approach (Model-A) from this work. Lindqvist [31] conducted the fatigue test for toe crack in a non-load-carrying cruciform weld joint having 0.6 mm toe radius. The reported results are compared with the current approach (Model-A) under tension mode only.

In this study, the new value of FAT95% for non-load carrying fillet weld having toe radius have been calculated equal to 71 MPa.

The new calculated FAT value is higher than that from IIW recommendations (FAT95%, 63 MPa) due to the effect of improved local weld geometries and stress concentration.

To verify the predicted value, the mean or design curve is drawn as shown in Figure 4.16. The mean curve fatigue life (FAT50%) calculated is based on FAT95%+2$STDV$. FAT50% is compared with reported test results from Lindqvist [31]. An initial crack length equal to 0.1 mm was adopted.

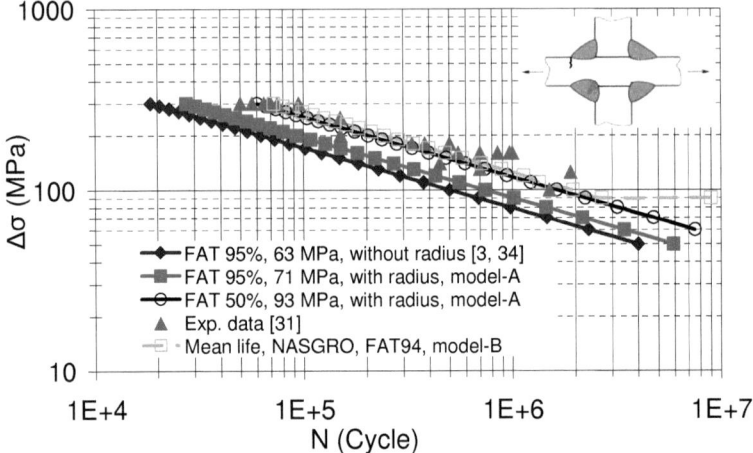

Figure 4.16: Calculated fatigue life compared with experimental results reported by Lindqvist [31]. Non-load-carrying cruciform joint with toe crack=0.1 mm, R=0 (C_{char}=5E-13, C_{mean}= 2.17E-13, m=3), NASGRO parameter are listed in Table 5.1 of Chapter 5, Steel S 403

The mean fatigue life (i.e., 50% failure probability, 50% survival probability) has been calculated equal to 93 MPa (FAT50%), as shown in Figure 4.16. It is shown that the calculated data are consistent with the experimental mean data corresponding to the weld joint of a small scale thickness joint (less than 20 mm). It means that the effect of thickness and residual stress through thickness was neglected. The calculated mean life using NASGRO equation (Eq. 4.9) shows a correspondence with the experimental data and FAT50% from Paris' law at stress ratio R=0.

The National Institute for Materials Science (NIMS) [56] presented a data sheet on fatigue properties of non-load-carrying cruciform welded joints of SM490B rolled steel for welded structures. The plate thickness 80 mm is investigated. Figure 4.17 shows the comparison between NIMS data and current calculations from this work (Model-A). The calculated mean fatigue life (FAT50%) has a good agreement with the experimental data. Also the experimental mean-life is in an agreement with NASGRO solution as shown in Figure 4.18 at R=0 and 0.1 mm toe crack.

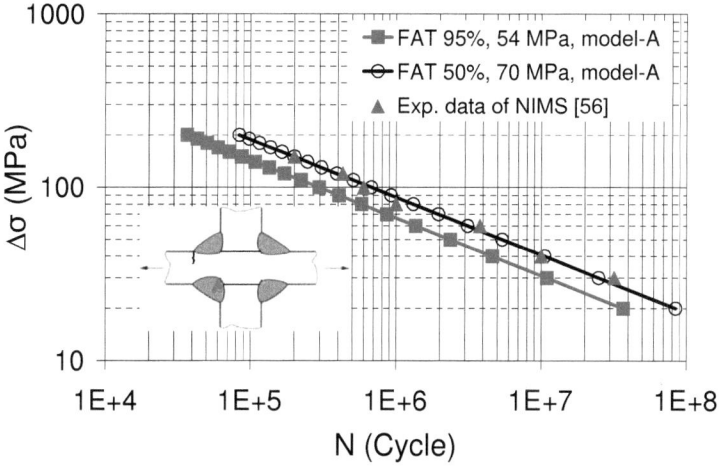

Figure 4.17: Calculated fatigue life compared with experimental data from NIMS [56] for non-load-carrying cruciform joint, thickness=80 mm (C_{char}=5E-13, C_{mean}= 2.17E-13, m=3)

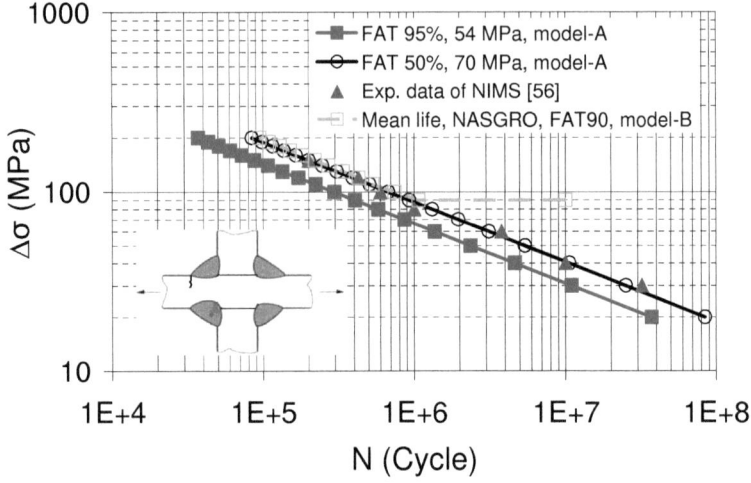

Figure 4.18: Comparison between NASGRO, fracture mechanics and experimental data from NIMS [56] for non-load-carrying cruciform joint, thickness=80 mm, R=0, (C_{char}=5E-13, C_{mean}= 2.17E-13, m=3), NASGRO parameter are listed in Table 5.1 of Chapter 5, Steel S 403

In case of LOP, the current approach (Model-A) is compared also with published test results by Singh et al. [84]. They carried out fatigue life experiments on gas tungsten arc welded load-carrying cruciform joints made of AISI 304L stainless steel in the cold rolled state of 6 mm thickness. Their experimental results were reported and identified as propagation and initiation life for LOP equal to 2, 3 and 6 mm. The FAT value increases as the LOP decreases as shown in Figure 4.19. For 2 mm, FAT equals to 74 MPa, whereas it equals to 61 MPa and 34 MPa in case of 3 mm and 6 mm LOP, respectively. The decrease in FAT strength as LOP increases is due to the decrease of the crack path to reach final length of failure and the fatigue life will thus decrease.

In spite of relatively high residual stresses which are likely to occur in the welds, several works proposed that residual stresses can be neglected or they have relieved. However, these considerations are a conservative assumption [29, 84]. NASGRO solution is developed at $R=0$ and compared with the current examples (see Figure 4.19). It is found that in case of LOP, NASGRO provides higher fatigue life because the large initial crack length has been used which is equal to LOP.

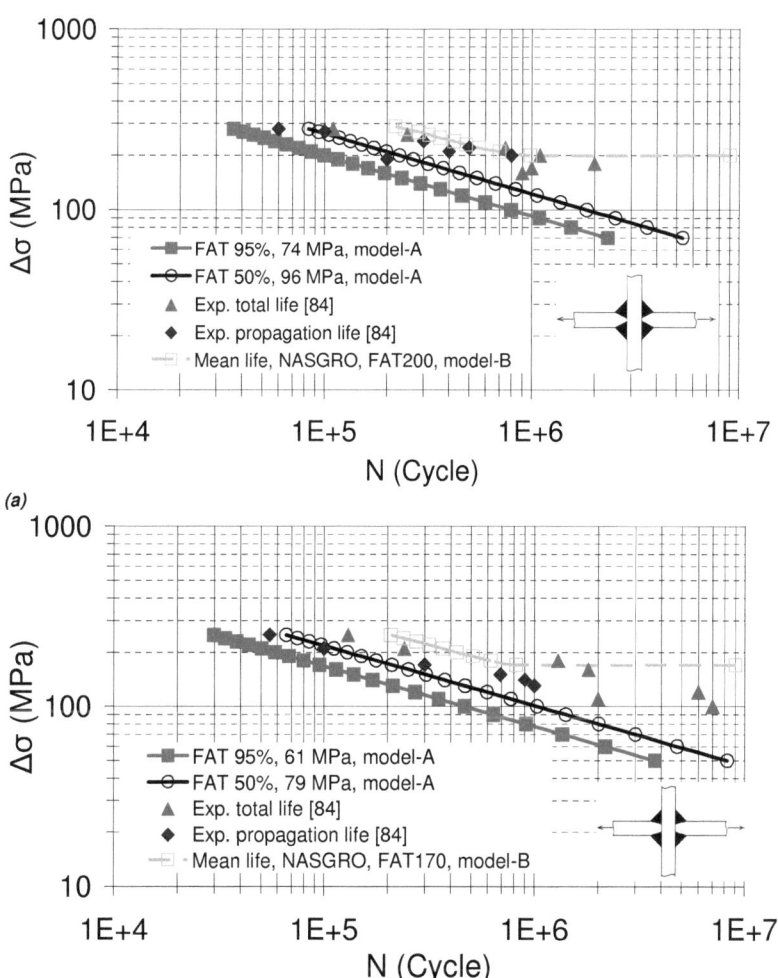

(b)

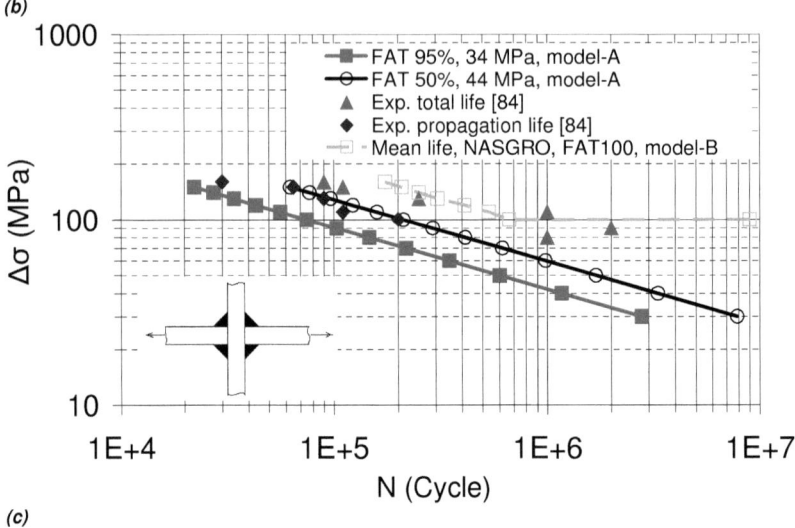

(c)

Figure 4.19: Calculated fatigue life based on Paris and NASGRO equations compared with experimental data reported by Singh [84] at R=0, (C_{char}=5E-13, C_{mean}= 2.17E-13, m=3): NASGRO parameter are listed in Table 5.1 of Chapter 5, Steel S 403(a) LOP= 2 mm; (b) LOP=3 mm; (c) LOP=6 mm

The fatigue lives for different LOP (2, 3 and 6 mm) are presented in Figure 4.20. The lower curve refers to FAT95% equal to 34 MPa in case of LOP=6 mm. It can be concluded that the LOP equal to 6 mm will be a maximum critical LOP, which can be allowed according to the weld size (throat thickness> plate thickness/3), where the FAT value equals 34 MPa issued in accordance to GL [13].

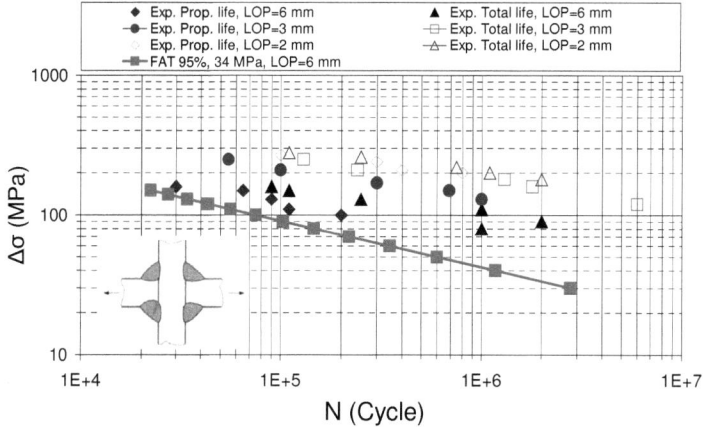

Figure 4.20: *Fatigue life for different LOP sizes published by Singh [84] compared with current calculated life based on fracture mechanics*

Singh et al. [84] calculated both lives (initiation and propagation life) and estimated that both initiation and propagation behaviour are equally important. Their results led to the crack initiation-propagation model (IP model) in which the total life N_T was the summation of initiation and propagation cycles, N_I and N_P, respectively. They calculated fatigue life N_T, which mainly based on the fatigue crack growth Eq. (4.1).

In LOP case, the initial crack length will be equal to LOP, and that is recommended in mind of the author since it shows good results compared with experimental data. Moreover, the considerable amount of scatter in fatigue life as compared with the experimental test data is small at high stress level. It is to be emphasized that the smaller scatter at high stress level is believed to result from a much shorter or negligible initiation period to crack propagation.

Therefore, a criticism could be put on the IP model which was used in previous works, because no significant initiation time is necessary.

Modelling description and calculated results are given in Table 4.4.

Chapter Four Fatigue Life Calculations and Verifications

Table 4.4: FE models, FAT, crack growth parameters, initial and final crack lengths

Case FAT 95%	Description	FAT50%	Crack growth parameters	Joint type and potentially crack locus (FE model)
217 45	Transverses butt weld with partial penetration, analysis based on stress in weld throat sectional area. Weld overfills not to be taken into account. **Geometrical conditions** Max. LOP≤0.2t but max.=2.2 mm.	59	a_i=1.1 mm C=2.17E-13 m=3	• LOP in transverse butt weld joint • FIX x and at corner FIX xy. • Weld reinforcement not to be taken into account
213 80	Transverse butt weld with weld toe crack and complete penetration. **Geometrical conditions** Toe angle max. 30°. Weld bead height max. =1+0.1×weld bead width. Crack will propagate through the sheet thickness.	105	a_i=0.1 mm C=2.17E-13 m=3	• Weld toe crack. • One side locked in x-direction. At least one point FIX xy to prevent the model from rotation.
414 36	Load-carrying cruciform joint with LOP. Incomplete penetration and un-weld line regards as initial crack (a_i), 0<a_i<sheet thickness. Throat thickness>t/3. **Geometrical conditions** Equal sheets thickness (B=T) and leg length (S=h). Throat thickness>t/3, a_f=Y× (leg length+thickness/2). Y: factor vary from 0.6-	47	a_i=7 mm C=2.17E-13 m=3	• LOP in load-carrying cruciform joint • FIX xy at corner. • Un-weld line was regards as a_i

414 40	Load-carrying cruciform joint with LOP. Throat thickness<t/3. Equal leg length (Isosceles). LOP propagates through the weld metal perpendicular to the applied load and coincide with the LOP line. **Geometrical conditions** Equal sheet thickness ($B=T$) and leg length ($S=h$).	53	a_i=4 mm C=2.17E-13 m=3	• LOP in load-carrying cruciform joint. • FIX xy at corner. • Un-weld line was regards as a_i.
413 63	Numerical analysis of 2-D load-carrying cruciform joints, fatigue crack propagation life is determined automatically for the 2-D models through integration of the Paris equation for a_i=0.1–a_f. **Geometrical conditions** Equal sheet thickness ($B=T$), leg length $S=h$, throat thickness <t/3.	83	a_i=0.1 mm C=2.17E-13 m=3	• Weld toe crack. • LOP is inactive.
412 71	Non-load carrying with toe crack. **Geometrical conditions** Equal sheet thickness ($B=T$), leg length $S=h$, throat thickness <t/3. Weld toe radius is ≤0.6 mm.	93	a_i=0.1 mm C=2.17E-13 m=3	• Weld toe crack. • LOP is inactive.

All cases have a transition radius of 1 mm, as suggested by different investigations [32].

In this work since there is no bending moment in above cases, the nominal stress is equal to the applied stress [23]. According to Table 4.4, the FAT-value at 95% survival probability seems to be multiplied with a factor 1.3 to achieve 50 % survival probability of failure. Therefore, we can obtain the following expression for FAT:

$$FAT_{char} = \sqrt[3]{\frac{C_{mean}}{C_{char}}} FAT_{mean} = 0.757 FAT_{mean} \qquad (4.19)$$

Nykänen et al. [15] also presented the following Eq. (4.20) to correlate the FAT-value with the mean life as follows:

$$FAT_{char} = \sqrt[3]{\frac{C_{mean}}{C_{char}}} FAT_{mean} = 0.8275 FAT_{mean} \qquad (4.20)$$

4.10. Effect of Residual Stresses

The residual stress has an important role on fatigue crack growth. Some researches have investigated the influence of welding residual stresses for crack growth analyses. The residual stresses might greatly influence the life or the geometry of the crack after growth. The influence has been investigated for through-wall cracks by some investigators, but there were only a few investigations for surface cracks, and the influence has not been revealed [99]. Usually, the effect of residual stresses and thickness are ignorable for sheet thickness less than 20 mm [20, 84].

The suggestion for the next Chapter Five is to investigate the influence of residual stresses on fatigue life curves for a surface crack analytically using the weight function which assumed a nonlinear stress field present through the thickness.

4.11. Effect of Sheet Thickness

FAT takes an increasing thickness of more than 25 mm into account with a reduction factor $f(t)$. As a complement to the testing, fracture mechanics calculations are made in an attempt to evaluate the effect of the stress concentration thickness dependence.

The stress concentration is defined as a local rise of stresses due to a geometrical change cross section reduction or discontinuity. A more or less abrupt change in geometry is often referred to as a notch.

Typical stress concentrations due to notches are shown in Figure 4.21. In this case, no reduction of the cross section is presented, but overfill of the weld metal will act as a stress riser. The stress will rise at the transition between the base plate and weld metal.

This area is often referred to as the weld toe. This is the potential crack initiation position [1].

The local geometry can be characterized by its flank angle θ and radius ρ. According to Figure 4.21, the stress concentration factor is defined as [1]:

$$K_t = \frac{\sigma_l}{\sigma_n} \tag{4.21}$$

where, σ_n is nominal stress which can be defined in the section away from the notch or the nominal stress in the notch section, σ_l the local notch stress.

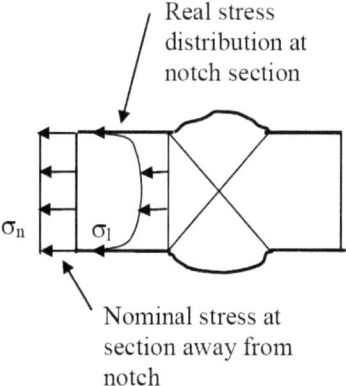

Figure 4.21: Stress concentrations at notches of butt joint [1]

The geometry of the item is important and is one of the issues that must be addressed at the design stage to achieve improved fatigue durability. The main issue is to reduce the stress concentration factor K_t in areas susceptible to fatigue [1].

To compile the geometric part of the thickness effects and validate the testing, a fracture mechanics analysis is carried out. Therefore, the plate thickness correction factor is not required in the case of assessment based on fracture mechanics procedures or effective notch stress [3, 34].

It is regarded that this neglecting of thickness' effect in fracture mechanics method return to the stress concentration phenomena. As the sheet thickness increases the stress concentration (K_t) will increase in local points where the crack has assumed and expected to initiate and propagate according to mechanisms of fracture mechanics method of materials.

From a fracture mechanics point of view the stress gradient effect can be understood because a crack at the surface of a thick plate will grow for a greater distance at a higher stress level as compared to a crack of the same length in a thin plate for the same surface stress. Thus, the thinner specimen will have a longer fatigue life [15]. Therefore, as the sheet thickness increases, the fatigue life strength (FAT) decreases without needs to the reduction factor as based on fracture mechanics point of view.

To verify this situation, one example of transverse butt weld joint was studied with FE simulation. Figure 4.22 shows the effect of sheet thickness on the stress distribution of transverse butt weld joints. The maximum stress at the weld toe decreases when the thickness increases from 10 mm to 40 mm at constant weld bead height and weld bead width equal to 2 mm and 10 mm, respectively.

Several iterations of the same model were analysed to select an appropriate mesh density and subdividing segments. The problems of missing elements and meshes commonly occur when using a fine mesh density, or inappropriate mesh ratio especially at the corners and edges. Anther problem is the crack increment when using a fine mesh, i.e., the crack is not able to reach its final length (one-half of the plate thickness). It is recommended to use larger increment for such problem in case it does not affect the SIF solution.

Moreover, the problem of memory spaces to solve certain cases will appear for a big geometry with large number of meshes, high meshes density at the corners, and multi subdivided areas.

Figure 4.23 and 4.24 show the effect of thickness on the local stress and stress concentration at the weld toe.

Chapter Four   Fatigue Life Calculations and Verifications

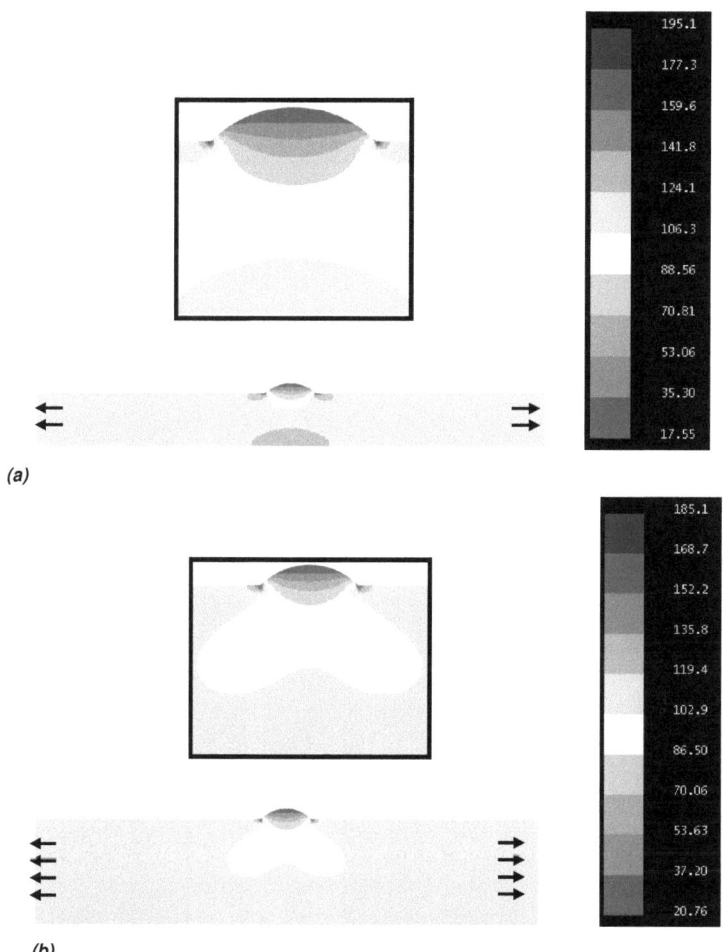

(a)

(b)

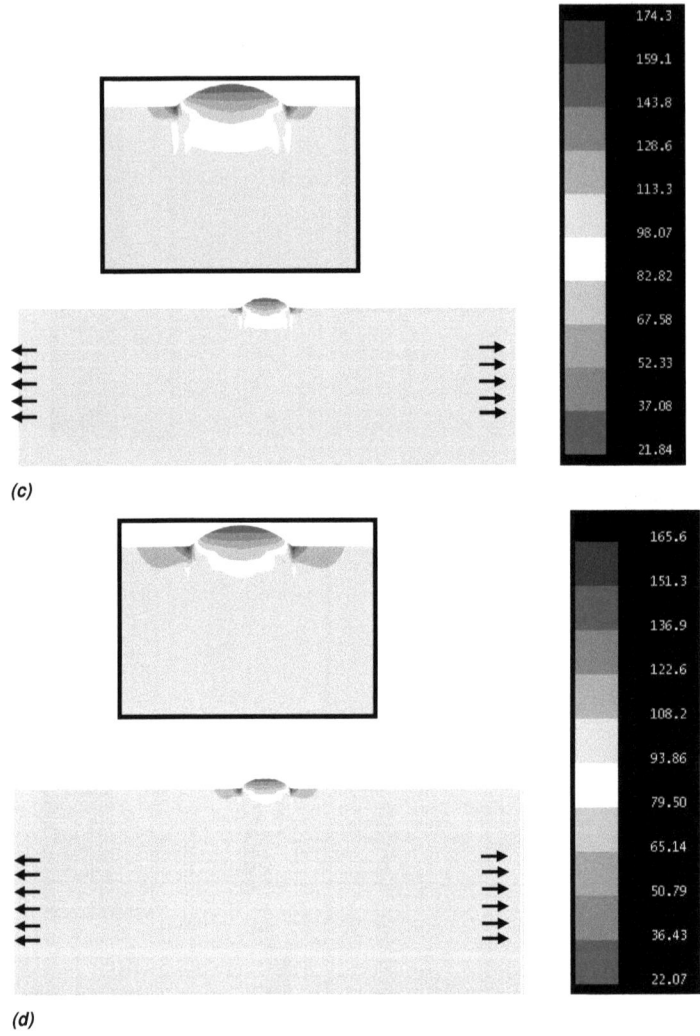

Figure 4.22: Screenshots from FRANC2D of the stress distribution (MPa) in V-groove butt weld joint at an external stress of $\Delta\sigma = 104$ MPa: The effect of thickness: (a) thickness=10 mm,

$\sigma_{max}=195$ MPa; (b) thickness=20 mm, $\sigma_{max}=185$ MPa; (c) thickness=30 mm, $\sigma_{max}=174$ MPa ; (d) thickness=40 mm, $\sigma_{max}=165$ MPa

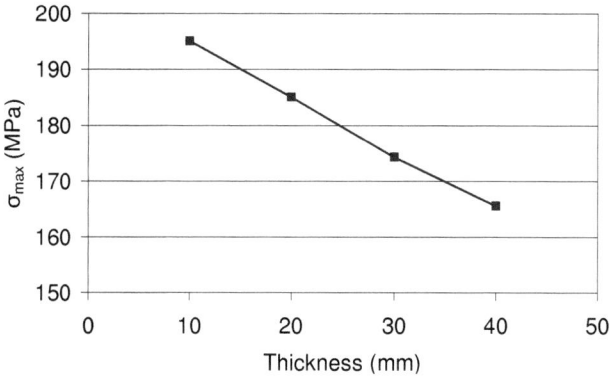

Figure 4.23: *The maximum stresses at weld toe for different thicknesses (MPa) in V-groove butt weld. Thicknesses are 10, 20, 30 and 40 mm for transverse butt welded joint*

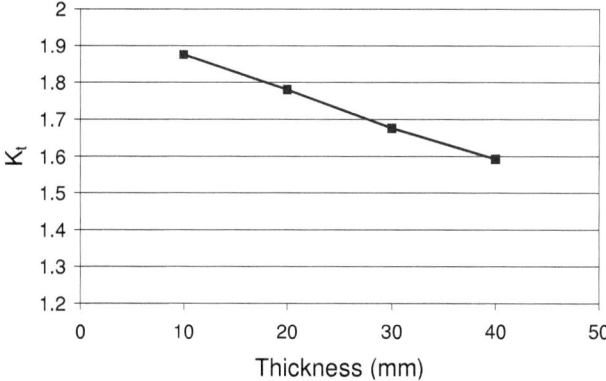

Figure 4.24: *The stress concentration factor for different thicknesses (MPa) in V-groove transverse butt welded joint*

It is to be emphasized that stress distributions and stress concentration effects in reality are changeable with respect to thickness, where the weld size and notch effect in combination with the welding conditions are different. Figure 4.25 shows the FAT for the four thicknesses of butt weld joints.

The comparisons of FAT-values are presented in Figure 4.25 for a given initial crack length of 0.1 mm.

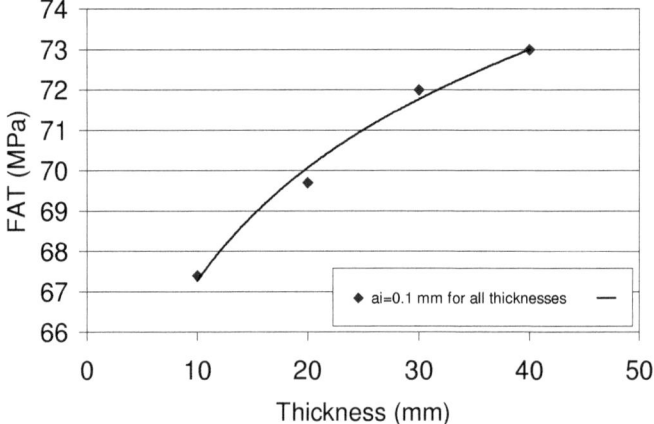

Figure 4.25*: FAT-values for different thicknesses (MPa) in V-groove transverse butt welded joint*

According to the IIW recommendations [3, 34] the reduced strength is taken into consideration by multiplying the fatigue class of the structural detail by the thickness reduction factor as follows [3, 31, 34]:

$$f(t) = \left(\frac{25}{t_{ref}}\right)^n \tag{4.22}$$

where $t>25$ mm, and n is the reduction factor exponent, n_{red} according to IIW recommendation (see Table 4.5)

The most common way to deal with the thickness effect is to use a correction factor that is calculated with the formula $(t_o/t)^n$ (Eq. 4.22). The correction factor is multiplied with the fatigue strength to achieve the reduced fatigue strength.

The reference thickness t_o is a subject of discussion; it usually varies between 15 - 32 mm in the different standards, recommendations and articles. The exponent n_{red} depends of the type of joint and typically varies between 0.07 and 0.3. The reference thickness t_o and n_{red} for different standards, recommendations and articles are listed in Ref. [31]. t_o is equal to 25 mm in IIW [3, 34] and Eurocode [12]. If $t < t_o$ it is usually recommended to choose $t = t_o$. No standards or recommendations consider the fact that there might be an increase in fatigue life for decreasing thickness below the limit value t_o [31]. The thickness correction exponent n_{red} depends on the effective thickness t_{eff} and the joint category as shown in Table 4.5:

Table 4.5: *The thickness correction exponent n for different weld types [3, 31, 34]*

Joint category	Condition	n_{red}
Cruciform joints, transverse T-joints, plates with transverse attachments	as-welded	0.3
Cruciform joints, transverse T-joints, plates with transverse attachments	toe ground	0.2
Transverse butt welds	as-welded	0.2
Butt welds ground flush, base material, longitudinal welds or attachment	any	0.1

Lindqvist [31] made two specimens with the geometry obtained from the real test specimens for two thicknesses, 6 mm and 12 mm. The initial crack was constant for both thicknesses equal to 0.15 mm. Finally he found that to get the calculation to be in accordance with the real fatigue testing, the initial crack must be set to be approximately 0.4-0.5 mm.

Therefore, a trial to find the effect of scaling crack length on FAT is made using the current approach. Figure 4.26 shows the FAT-values calculated from IIW and current work in case of using proportional scale crack lengths, 0.1, 0.2, 0.3, and 0.4 mm for thicknesses 10, 20, 30, and 40 mm, respectively. The proportional crack length would give expected results of FAT.

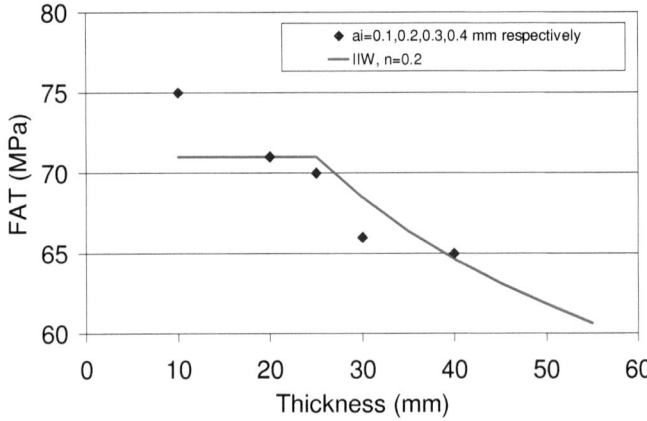

Figure 4.26: *Effect of thickness using the reduction factor from IIW (equation 4.22 and FAT71) compared with current work using the proportional scaling of crack length*

4.12. Geometrical Verification of Butt Weld Joints

The fatigue life assessments of butt welded joints are carried out frequently in different experimental tests. Therefore, in these following parts, the effect of weld geometry of transverse butt welded joints will be discussed according to the current fracture mechanics approach and is based on the calculated crack parameters from this work.

4.12.1. Simulation of Fatigue Crack Growth

Finite Element Method (FRANC2D software) has been used to find an appropriate solution of SIF. It is found that SIF obtained by FEM gives better crack growth direction and solutions for different notch geometries as compared with analytical formulas, also the current FE models are in agreement with the experimental results from literature.

From the previous work [100], rules have been given for estimation of the crack growth direction, initial crack length and growth parameters, C and m. The value of C and m in case of characteristic values of FAT 95% (5% failure probability) are determined equal to 5E–13 and 3, respectively. FAT 50% (50% failure probability) are determined according to C50% and m equal to 2.17E–13 and 3, respectively. The FEA is used to simulate fatigue crack growth. As a fatigue crack propagates, the re-meshing process will be carried out automatically using FRANC2D program.

Figure 4.27 presents the simulated and real experimentally tested specimens, where W is the reinforcement width (weld bead width) and H the reinforcement height (weld bead height). Fatigue crack growth is simulated under the opening mode-I which is usually assumed in fracture mechanics.

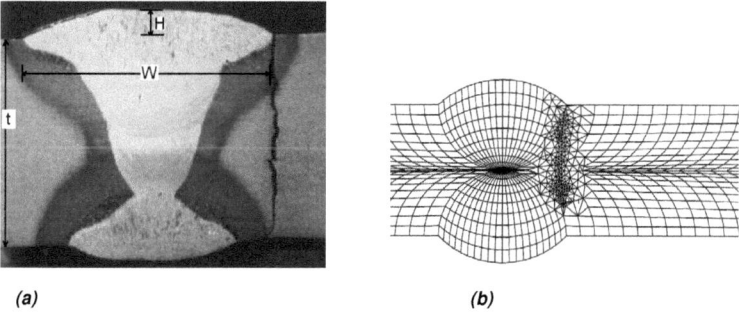

(a)   *(b)*

Figure 4.27: Transverse butt weld joints: (a) experimental crack [101]; (b) crack growth simulation by FRANC2D

It has been shown that fatigue crack growth behaviors of welded joints are highly dependent not only on the materials and load conditions, but also on weld geometry such as weld toe angle, weld toe radius, plate thickness and width of reinforcements (the weld bead) [28]. Nguyen et al. [50] used LEFM to present the effects of tip radius of undercut at weld toe, weld toe radius, flank angle, plate thickness and edge preparation angle on the fatigue crack propagation life. In addition, the current work highlights on the weld bead height, weld bead width and on sheet thickness effects which are regarded the main factors which connect the other remaining parameters. Cracks occur at the weld toe where high residual tensile stresses are introduced by the welding processes due to the expansion and the shrinkage of weldment during heating and cooling, misalignment and microstructural variation in weldment and HAZ, see Figure 4.27.

4.12.2. Failure Mode

Due to symmetry, the half models can be used to determine the fatigue propagation life under cyclic tensile loading, as described in Chapter Three. The majority of crack initiation is at geometrical discontinuities such as weld toe and weld defects. The maximum depths at the fatigue crack initiation points are found to be less than 0.016 in., or 0.4 mm. While the maximum distance from the weld surface to the embedded defects is about 0.08 in., or 2 mm [22, 102]. Therefore, it can be supposed that the initial crack length has a range between 0.4-2 mm. The uncertainly of initial crack lengths is incredible and leads to serious results; therefore, for all the models, an initial crack depth of weld toe crack of 0.1 mm is used as predicted before.

The FEM simulation by FRANC2D shows that the maximum stress intensities develop when the plane of the crack is normal to the direction of the primary tensile stress (opening mode). However, as the plane of the crack tilts further away from the mode-I tensile loading, the structural member, in the form of a plate or a similar configuration, will lose its ability to support the external load.

4.12.3. Weld Metallurgy and Defects

It has been generally recognized that welded structures contain defects, either built in during fabrication or initiated early under service conditions. When the structures are subjected to variable cyclic stress, fatigue cracks are initiated from these defects and propagated from a sub-critical to a critical size. These cracks are usually initiated at the weld toe. Therefore, welds are prime target in most failures. Subtle peculiarities lie at the root of many weld designs. These are related to the heat-affected zone (HAZ), weld proper and the fusion line. The fusion line seldom propagates fracture unless it becomes very brittle. More description of several principal type of cracking that can occur during or after welding can be found elsewhere [103].

Impurities consisting of a small amount of sulfur and phosphorus are critical to hot cracking of steel which mechanism of cracking it appears at high temperature before the metal develops sufficient strength, see Figure 4.28. Cold cracking occurs after hard products such as bainite and martensites are formed [103].

The main cause can be treated to inherent metallurgical sensitivities and un-proper weldment. Figure 4.28 shows some of the traditional weld defects.

For welded joints, the presence of weld imperfections such as slag inclusions at weld toes, undercut, residual stresses, lack of penetration LOP etc., effectively will reduce the initiation phase. All welded joints contain small slag inclusions at the weld toe, which acts as pre-existing cracks and stress-raisers. Fatigue crack propagation commences from these inclusions very early in the life [8]. By contrast to fracture mechanics, cracks are assumed in those regions according to Figure 4.28.

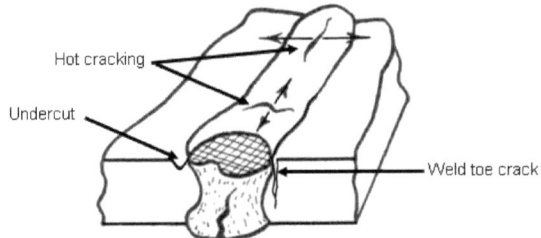

Figure 4.28: Butt weld defects; toe crack, undercut condition and hot cracking, after [103]

Nevertheless, other defects can be found due to non proper weld technique and fabrication faults. Therefore, most researchers and engineers used to assume that there are crack-like defects in welds [39]. Then the total fatigue life of welded joints N_T is the stress cycles devoted to the crack growth from these defects to final failure i.e., $N_T=N_P$.

Lindqvist [31] showed a 12 mm specimen's fracture surface after it was fatigue tested and afterwards broken up. Probably there were several small cracks along the weld toe. When the cracks grew in the direction normal to the applied load they united into one large semi-elliptical crack in direction normal to applied load. These behaviours are quite similar for most joints having toe cracks.

4.12.4. Fatigue Life and Crack Growth

Numerical analysis of 2-dimensional, complete penetration butt joints is performed to determine the effect of weld profile and geometries on fatigue crack propagation life under cyclic tensile loading.

Many applications in the structural areas involve welded components, which have to be designed to avoid fatigue failure. Though considerable fatigue data are existent for welded joints in structural steels and aluminum in IIW and BSI, there are very sparse fatigue life design data for welded joints.

Chapter Four   Fatigue Life Calculations and Verifications

The recommended standard doesn't include the effect of different geometries, for example regarding the toe radius; BS doesn't regard the weld toe radius is constant. Therefore, the assessment of welded joints is a major industrial problem, for three reasons. Firstly, welds tend to be regions of weakness in a structure due to stress concentration effects and poor material properties. Secondly, it is difficult to predict their behavior accurately. This is partly due to the difficulty of defining material properties, which vary throughout the weld and HAZ, but a more crucial problem is the difficulty of defining weld geometry in a manner which is sufficiently precise for analysis and sufficiently simple for industrial use. Moreover, geometry has a major influence on fatigue strength producing scatter of results which leads to uncertainties about the magnitude of cyclic stress, which can be applied to a component for which a particular fatigue life is required and statistics are normally used to interpret test data.

The third reason is regarding the time that fatigue testing machines, particularly those needed to test relatively large specimen, apply cyclic loads at frequencies typically in the range 5-15 Hz. Clearly, this means that the generation of fatigue test data can be a lengthy process, even for one type of specimen, and one particular stress ratio. Indeed, it takes 8-24 days to apply 10^7 cycles, for this most test series are normally confined to fatigue lives of less than 10^7 [104]. Therefore, fatigue life determination of welded joints is a quite complex problem. All theories and models have to be verified and corroborated by experimental data. Hence, the advantages of the present's work approaches are very evident. Traditionally, the linear elastic fracture mechanics (LEFM) approach which estimates the crack propagation life N_P was used to calculate the total life N_T of welded joints. However, the LEFM approach always gave conservative results of N_T when compared with those of modern experimental data of weldments [84].

4.12.5. Modeling

The determination of the SIF for the 2-dimensional butt weld joints is carried out using LEFM analysis. This method is well encoded in FRANC2D; the FE package is used in this study.

The material type used for the base and weld metal is steel, so values of E are chosen as 210 GPa. Figure 4.29 shows the FE models that are used in this study with approximated geometries for the total joint length equal to 10 cm.

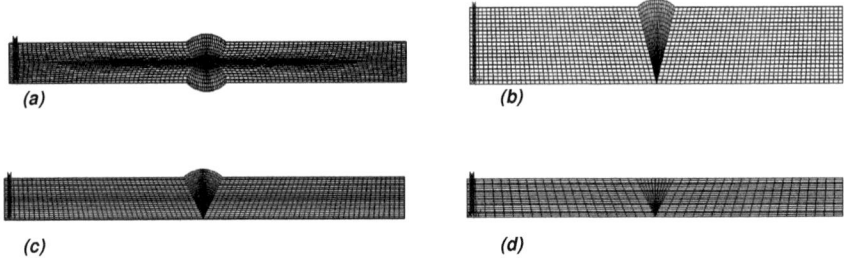

Figure 4.29: Meshes of different geometries and thicknesses of butt weld joints: (a) unmachined double sides X-groove; (b) unmachined V-groove; (c); unmachined and; (d) machined V-groove for thinner plate

4.12.6. Verification Results

Butt welds with different thicknesses and under the effect of weld reinforcement width and height are presented to show the validity of the current approach (Model-A) to calculate fatigue life under the effect of geometry.

Transverse butt weld made in shop in flat position with toe angle more than $30°$, as in Figure 4.29(a), has FAT equal 80 MPa from IIW case 213 [3, 34] and GL, type No. 3 [13]. Figure 4.30 shows the fatigue life calculation for this case using numerical integration of Paris' law. Initial crack length equal to 0.1 mm is confirmed with the FAT value from IIW and GL. Therefore, it can be concluded to use this crack length for other cases of butt welds, single or double side welds.

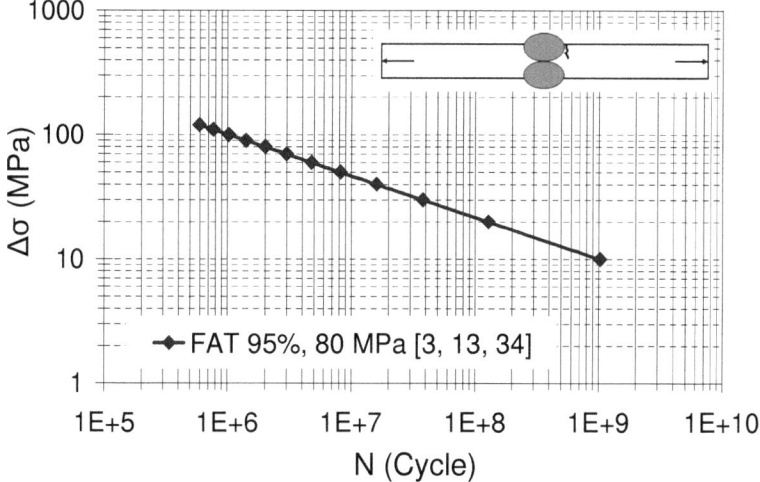

Figure 4.30: S-N curve for X-groove un-machined butt weld, weld reinforcement width, height and plate thickness are 10 mm, 2 mm and 10 mm, respectively, (C_{char}=5E-13, m=3)

The S-N curve for single groove butt weld with sheet thickness 20 mm, weld reinforcement width and height 10 mm and 2 mm, respectively (see Figure 4.29b), are presented in Figure 4.31. This calculated FAT value coincides with those recommended from GL type No. 6, FAT71 [13].

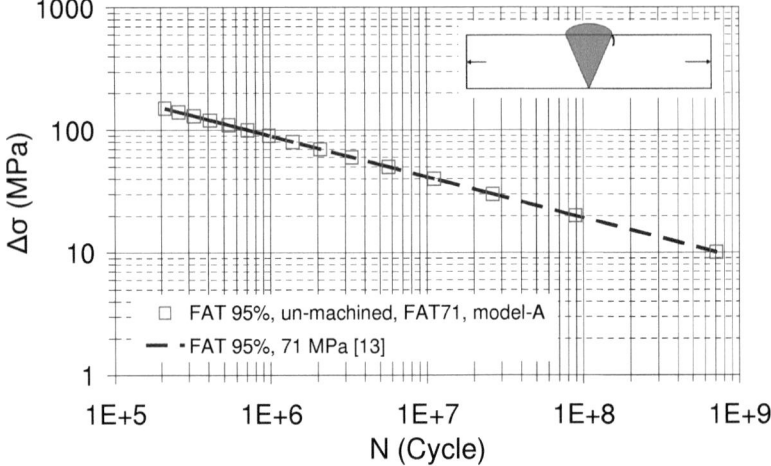

Figure 4.31: S-N curve for V-groove un-machined butt weld, weld reinforcement width, height and plate thickness are 10 mm, 2 mm and 20 mm, respectively, (C_{char}=5E-13, m=3)

The stress concentration is increased as a local stress is raised due to the geometrical change and the weld overfilling.

Overfill of the weld metal will act as a stress riser. This stress rise will of course be very harmful with respect to fatigue damage. Therefore, and due to the effect of stress concentration, the fatigue strength decreases with the presence of weld reinforcement. Hence, the complete removal of the weld bead reinforcement (see Figure 4.29d) is reported to give improved fatigue performance in condition of free surface from crack-like scratch after machining. The machining processes for fillet weld is highly effecting on FAT and fatigue behavior of joints. In this work, the flash grinding increases the calculated FAT to 94 MPa while it was 71 MPa as reported in Ref. [13] in case with weld reinforcement due to reduction of stress concentration at weld toe, see Figure 4.32. An increase in endurance limit is also observed in the joint area of flash butt welded joints after the weld flash has been removed [105].

It is to be emphasized that in case of machined overfill, the model-A using the initial crack length of 0.1 mm shows a lower FAT, because the machined joint in reality has a lowest initial crack length.

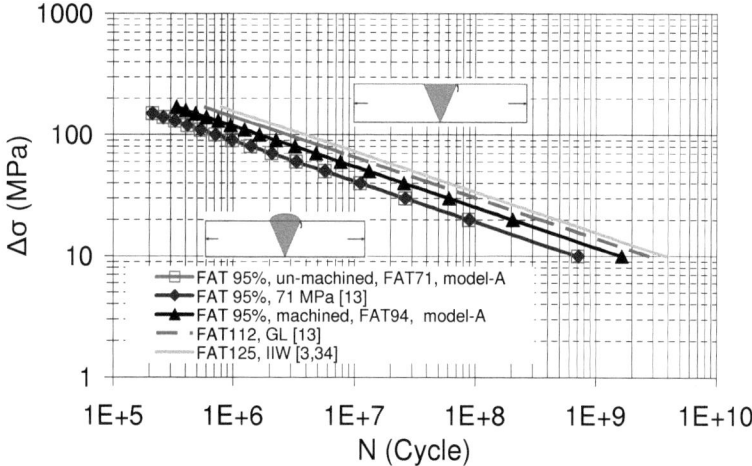

Figure 4.32: S-N curves for V-groove un-machined and machined butt weld, weld reinforcement width, height and plate thickness are 10, 2 and 20 mm, respectively, (C_{char}=5E-13, m=3)

Figure 4.32 shows that the decrease in FAT between smooth plates (machined butt weld) and notch plate (unmachined butt weld) can be defined in terms of the stress concentrating factor as follows:

$$K_t = \frac{FAT_{machined}}{FAT_{unmachined}} \quad (4.23)$$

FAT-values for machined transverse butt weld joint are equal to about 94 MPa (see Figure 4.32), then, the K_t is equal to 1.33. In addition to IIW, GL [13] presented the values of FAT for machined and unmachined butt weld equal to 112 MPa and 71 MPa, respectively, where FAT112 is for v-groove and double sided weld, then K_f=1.58.

The comparison between 10 mm, and 20 mm thicknesses with constant a_i=0.1 mm is shown in Figure 4.33.

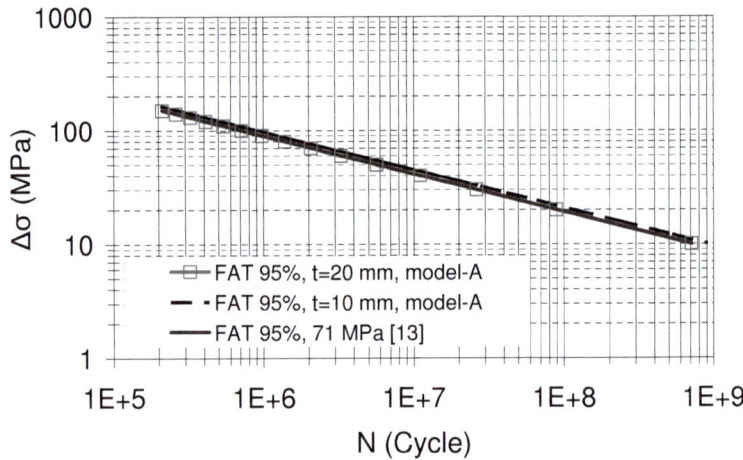

Figure 4.33: S-N curves for V-groove unmachined butt weld, weld reinforcement width, and height are 10, 2 respectively for the two thicknesses 10 and 20 mm, respectively, (a_i=0.1 mm, C_{char}=5E-13, m=3)

The comparison between different thicknesses (10, 20, 30, and 40 mm) with proportional scale of crack length is shown in Figure 4.34.

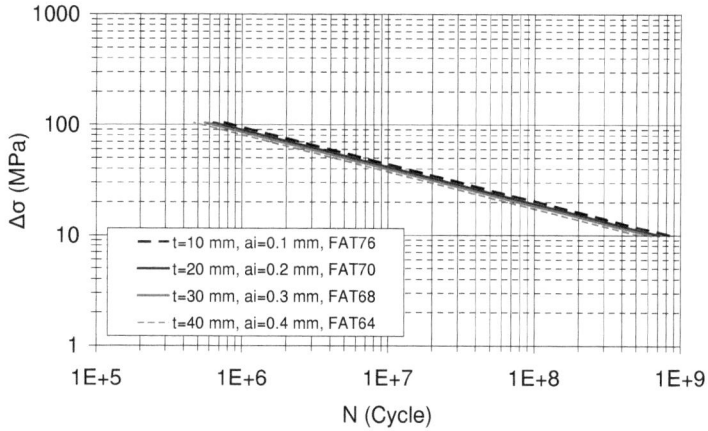

Figure 4.33: Characteristic S-N curves for V-groove unmachined butt weld, weld reinforcement width, and height are 10, 2 respectively for the thicknesses between 10 mm and 40 mm, respectively, (C_{char}=5E-13, m=3)

4.12.7. Standards Verifications

The predicted scatter band of *S-N* curves subject to the variation of all the weld geometrical parameters is in a good agreement with *S-N* curves covered by design classes from BSI 5400 [75]. The design class F is recommended to be used commonly for design of butt welded joints of the predicted scatter band of *S-N* curves in Ref. [50].

For reference, the design life and mean life of relevant full penetration arc welded butt joint fatigue classes as specified in BSI 7608 [76] are also shown. Class D welds are shop welds made in flat position using specific arc welding processes. Class E welds are out of position welds made by other processes including submerged arc welding. Class F welds are welds made on a permanent backing strip, and certain classes of fillet welds. Therefore, a high quality welded butt joint would be expected to be to class D [106]. Figure 4.34 shows the comparison with different design data of fatigue test for butt weld.

According to the current work, the fracture mechanics approach has been given more conservative fatigue strength, as compared with BSI 7608. Therefore, it is suggested that for safe fatigue design of critical welded practice, fracture mechanics should be used with crack parameters which are calculated in this work.

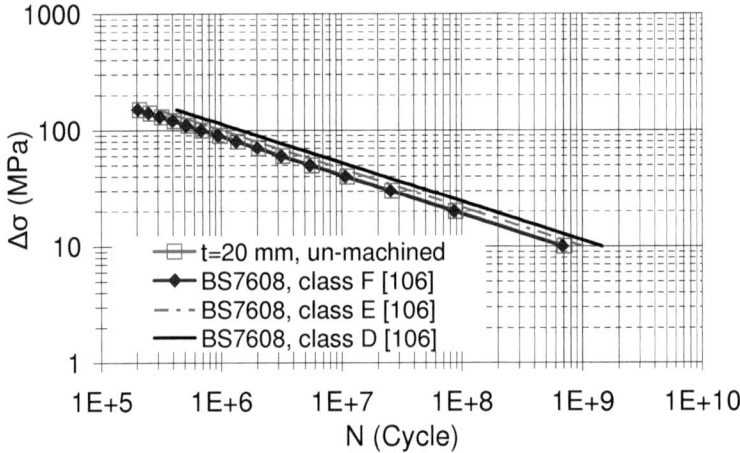

Figure 4.34: Comparison between different fatigue design data (FAT95%) and current calculation ($a_i=0.1$ mm, $C_{char}=5E-13$, $m=3$)

The current results agree well with literature which used LEFM and FEA and those based on experimental test [8, 28, 50, 67, 68, 72].

Butt weld joints have some defects during fabrication processes. Regardless to these defects, fracture mechanics which are used to predict fatigue life for those joints supposed that cracks already exist. The calculations of fatigue life are based on numerical integration of simple Paris' law and reliable solution of SIF, a_i, C and m. The initial crack length, a_i equal to 0.1 mm in case of weld toe is satisfying for different butt joints geometries. The machining of weld reinforcement will increase the fatigue life, both for single and double groove welds. Meanwhile the single groove weld has higher fatigue strength than the double side weld. The increasing of plate thickness will decrease the fatigue strength and the number of cycles to failure. Agreements with the recommendations and the experimental literature are obtained which validated the current calculation (Model-A) of initial crack length and convergence study of numerical integration. Hence, that prevents wasted time to carry out experiments. Therefore, it is suggested that for safe fatigue design of critical welded practice, fracture mechanics should be used with crack parameters which were calculated in this work.

4.13. Conclusions

Fracture mechanics is used to find the accurate prediction of fatigue life of welded joints. It can be used to determine FAT of unknown notch cases of welded connections. In the literature, different values of cracks length are proposed and presented normally as a range. Initial cracks take on the responsibility of massive break down. The main purpose for this study is to determine the crack lengths with the growth parameters. Some comparisons are carried out with experimental results to detect and identify fatigue life for more than one type of welded joints and type of cracks. Moreover, the FAT values are determined, which are a link to specific welded joints.

The entire fatigue process in fillet welded joints is modelled by pure fracture mechanics approach. The simple version of Paris' law has been adopted. In this work, the initial crack depth and growth rate parameters are determined according to backward calculations to determine FAT. The new values of FAT are calculated according to the predicted existence of an initial crack. Therefore, the proposed initial crack length is expected to give good results. An initial crack size equal to 0.1 mm is used for all joints that have weld toe crack, and this length is typical when arc welding is used and consistent for different welding processes. The different thicknesses are simulated for transverse butt weld joints and the effects on stress concentration and fatigue life are presented. It was found that for the constant initial crack, the FAT will increase as thickness increases due to reduction of stress concentration factor. However this conclusion is conflicted with the fracture mechanics point of view, but it is consistent with FE point of view in terms of constant weld geometry and nominal applied load. The scaling crack length with respect to plate thickness would give the expected results. The effect of scaling crack length on FAT-curves is also presented.

The root crack is varied depending on the degree of the penetration. The conventional crack lengths for joints having LOP or incomplete melting welds metal will be equal to line of LOP. These initial crack length values are applicable for all types of joints which have the same crack type. The cruciform fillet joints with LOP=2 mm show a better FAT value compared with longer LOP. The FAT value increases as the LOP decreases due to the increasing crack path to reach final length of failure.

The possibilities for simulating different geometries and calculating the fatigue life for welded joints that provide the fatigue strength are shown. These possibilities are not yet listed in previous recommendations. Thus save the time and costs that are needed for experimental testes. The final crack length has little effect as compared with the effect of initial crack. Therefore, it is defined to be equal to one-half of the sheet thickness in case of weld toe when the crack path is perpendicular to the applied load. Some other empirical equations were used for final crack length in case of LOP. Moreover, final crack length assumptions are verified from a-N curve, when the number of cycles becomes constant.

Chapter Five
RESIDUAL STRESSES IN WELDED JOINTS

5.1. Introduction

The effect of a combination of crack closure and residual stress distributions during the crack propagation has been investigated rarely.

In general, the modern standards and codes of fatigue design present data corresponding to the fatigue strength (FAT) of real welded joints. These data include the effect of welding technology, type of welded component and welding residual stresses. Nevertheless, in many cases, there is a need to consider the influence of welding residual stresses on the fatigue life of structural components in greater details.

The residual stresses have an important role on fatigue crack growth. Some studies [5, 65, 95] have investigated the influence of welding residual stresses on crack growth. The residual stresses might greatly influence the whole fatigue life. The influence of residual stress has been investigated for through-wall cracks by some researchers, but there were only a few investigations for surface cracks, and the influence has not been revealed yet [99].

In this work, the effect of residual stresses during crack growth of an edge crack is presented. The effect of residual stresses and geometry on the fatigue crack growth life of welded joints is studied by developing an analytical model using Linear Elastic Fracture Mechanics (LEFM), superposition method and FE approaches. Various configurations of residual stresses in welded joints are considered and the corresponding residual SIFs are calculated by using weight functions methods. The weight function is an analytical method which is used due to the nonlinear stress fields through the plate thickness.

Chapter Five   *Residual Stresses in Welded Joints*

Fatigue life of welds subjected to residual stress configurations in as welded condition are calculated by using the NASGRO equation with the total range of SIF (K_T) for different applied loading stress ratio (R), and compared with the available experimental data. The current procedure (Model-C) is verified with results found in the literature.

According to the developed approach from the current study, the effects of residual stresses and weld joint geometry on the fatigue behaviour of welded joints have been estimated.

In this chapter the T- and cruciform welded joints with weld toe crack are presented. In addition, Nykänen et al. [15] presented the fatigue life of transverse butt weld with LOP. This joint is used in addition to verify and compare the current residual stresses calculation.

5.2. Assumptions in Finite Element Model

The assumptions used in fracture mechanics are:
1. The material is homogeneous, isotropic and linearly elastic.
2. The crack tip is subjected to mode-I where only axial load is applied and mode-II and III where the biaxial load is applied.
3. The effects of plasticity ahead of the crack tip and in its wake are ignored as assumed by Alam [8].

In this chapter the effects of plasticity in terms of crack closure will be studied within the effect of residual stresses.

That "fatigue crack closure" is caused by the effects of local compressive residual stresses that are naturally developed by the fatigue crack growth mechanisms. On the other hand, the residual stress field due to welding "already" exists before a fatigue crack grows through this field (in fact these stresses change as a result of crack growth).

Therefore, the crack experiences a combined effect of a "developing" residual stress field (due to crack closure) and a "changing" residual stress field (due to welding) in a quite complex combination. These combinations are considered to be restricted

Chapter Five   Residual Stresses in Welded Joints

aspects and complex in several studies. Therefore, the inclusion of the effects of residual stress distributions, sign and direction are the most interesting challenges. These effects will be evaluated in this chapter.

5.3. Residual Stresses Assessment

Special conservatism is adapted to the fatigue strength to reduce the risk of unexpected failure due to geometry and residual stress effects.

Due to the high temperature gradients and plastic deformation in the vicinity of the weld during the welding processes, residual stress fields are invariably set up in the welded joints.

Usually such stresses are accounted for in safety assessment procedures such as the British R6 [107] and BS7910 [14]. Since the informations on residual stress distributions, signs and magnitude in the component are often not directly available, compendia with recommended upper-bound residual stress profiles for use in analyses are included in structural integrity assessment procedures for European industry (SINTAP) [69], British R6, and BSI 7910. These recommended profiles are classified according to the welding information.

On the other hand residual stress distributions can be measured experimentally by using different tests such as neutron or x-ray diffraction or they are estimated by using FEM.

Lee et al. [71] determined the residual stress distributions for plate T-butt welds from a detailed FEA of the welding process and these distributions were compared with those of the measured data for validation and shown that they have similar profiles. Lee et al. [71] also determined the linear elastic SIFs using the FEM and examined a new residual stress distribution which was less restrictive but had lower conservatism when compared with the analysis results from the measured data and the distributions of R6 and BSI 7910. The profile is applicable to different geometries of T-plate and tubular T-joints as well as different steels (Grades S355 and SE702).

Hyeong et al. [61] presented SIFs in welded T-plate and tubular geometries using measured residual stress distributions, and these are recommended in assessment procedures. Residual stress distributions were measured along the thickness direction (crack length up to half of the plate or pipe thicknesses, $a/t=0.5$) from the weld toe, where cracks are often found to initiate. The neutron diffraction method was used and stresses were measured along the thickness direction. A medium strength steel (BS EN 10025 Grade S355) and a high-strength steel (designated SE702, equivalent to the A517 Grade Q steel) have been examined for the T-plate. The former has yield strength of 348 MPa, and the latter has a yield strength of 700 MPa.

O'Dowd et al. [60] used the neutron diffraction method to determine residual stress distributions for welded T-plate. These distributions have been measured using neutron diffraction method which based on measured elastic strains to determine directly the stress field. They also examined a high-strength steel (700 MPa yield strength) and medium strength steel (348 MPa yield strength). The welding conditions and procedures were presented in details in their work. The resultant linear elastic SIF for these measured residual stress distributions, K_{res} has been obtained using FEM software package ABAQUS.

Researches up-to-date depend on FEM to find residual stress distributions and to calculate SIF including the effect of residual stresses K_{res}.

Therefore, the sensitive information needed to predict the behaviour of the welded structures are the residual stress distributions, $\sigma_{res}(x)$ and residual stress intensity factor K_{res}, where x is the vector along the crack path for each crack length increment.

K_{res} was calculated in the current work by the product of the weight function, $m(x,a)$ and the crack surface stress distributions, $\sigma_{res}(x)$. Moreover, in this work, the residual stress distribution is established by using SINTAP safety assessment procedures assuming the case of unknown welding conditions. These stress distributions are compared with experimentally measured data from literature.

In this work, the calculated residual SIFs are compared with the reported data for welded T-plate from Refs. [60, 61] for medium and high-strength steels. The transverse residual stress distributions were studied in this work dealing with mode-I fracture, and different residual stress distribution profile curves are presented using the SINTAP profile equation.

In addition, SIFs have been determined using FEM with the help of fracture mechanics which are encoded in the FRANC2D program. K_{res} is calculated by applying point load along the expected direction line of a toe crack. The latter calculation is less conservative and it has good agreement with K_{res} calculated from experimentally measured data is observed.

5.4. Residual Stress Intensity Factor

Residual stress intensity factors are necessary for the fatigue life and crack growth prediction. Therefore, the supposed weight function method is used. The weight function method is an analytical technique for deriving SIF from knowledge of the residual stress distribution in the un-cracked body.

The range of applicability of the weight function in Ref. [53] is restricted to somewhat limited weld geometries and the T-plate of Figure 5.1 falls outside this range. Therefore, O 'Dowd et al. [60] used FEM to determine the weld residual SIF for cracks of different sizes starting at the weld toe to overcome the use of the weight function. Also, they determined for comparison the SIFs for the stress distributions provided in R6 and BSI7910.

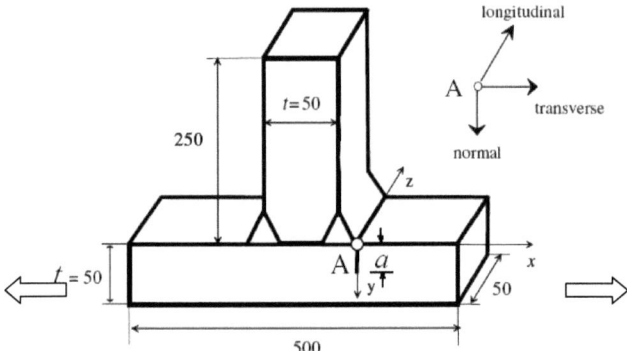

Figure 5.1: Geometry of T-plate and weld stress directions, all dimensions in mm [60, 61]

In the previous studies the SIFs were determined to the problem of a crack originating from an angular corner (weld toe) in the medium strength T-plate using the weight function for a T-plate geometry proposed in Ref. [53].

The weight function method is used to calculate the SIF because the exact solutions are not always available due to the complexity in either geometry or loading conditions [41, 53, 65]. Therefore, this method is often used because it enables the SIF for a variety of loading conditions to be calculated by integrating the product of weight function $m(x, a)$ and the crack surface stress distribution $\sigma(x)$ expression [41, 61], as described in Eq. (5.1):

$$K_{res} = \int_{x=0}^{x=a} \sigma_{res} m(x,a) dx \tag{5.1}$$

Because the SIF from Glinka and Niu's [53] weight function is only valid for the relative depth $a/t \leq 0.5$ [53], Gue et al. [41] derived widely applicable weight function solutions with a crack of relative depth a/t up to 0.8 originating from the weld toe. They used the $\sigma_{res}(x)$ distribution from FEA.

The numerical integration of Eq. (5.1) was carried out for the current T-plate welded joints as presented in Figure 5.1 and compared with Refs. [60, 61].

In general, the weight function method is widely used especially for cracked bodies with simple geometries. The advantage of the weight function method is that it enables the calculation of the SIFs in a loading-independent way because the weight function depends only on the geometry and boundary conditions, and is independent of the applied load. Still, there is no weight function method applicable to a general structure [41, 61].

More recent researches [41, 73, 74] have found that the following general weight function expression can be used to approximate weight functions for a variety of geometrical crack configurations subjected to one-dimensional mode-I stress fields [41, 53, 73, 74]:

$$m(x,a) = \frac{2}{\sqrt{2\pi(a-x)}}\left[1 + M_1\left(1-\frac{x}{a}\right)^{1/2} + M_2\left(1-\frac{x}{a}\right) + M_3\left(1-\frac{x}{a}\right)^{3/2}\right] \quad (5.2)$$

where a is the crack length and x the distance along the face of crack. M_i ($i=1, 2, 3$) are parameters which depend only on the geometrical configuration of the cracked body. The further details of the M_i determination can be found elsewhere [41, 73, 74]. The geometry factors, M_1, M_2 and M_3, for edge and through cracks in a finite width plate can be found as follows [41]:

$$\begin{aligned}M_1 &= \sqrt{2\pi}(3Y_1 - Y_u) - 4.8 \\ M_2 &= 3 \\ M_3 &= 3\sqrt{2\pi}(Y_u - 2Y_1) + 1.6\end{aligned} \quad (5.3)$$

with,

$$Y_u = 0.8843 + 4.3274\left(\frac{a}{t}\right) - 39.4056\left(\frac{a}{t}\right)^2 + 284.5721\left(\frac{a}{t}\right)^3 - 1038.1899\left(\frac{a}{t}\right)^4$$
$$+ 2116.4717\left(\frac{a}{t}\right)^5 - 2218.4035\left(\frac{a}{t}\right)^6 + 955.5433\left(\frac{a}{t}\right)^7$$
(5.4)

and,

$$Y_1 = 0.5854 + 1.8116\left(\frac{a}{t}\right) - 16.4166\left(\frac{a}{t}\right)^2 + 116.5429\left(\frac{a}{t}\right)^3 - 421.5994\left(\frac{a}{t}\right)^4$$
$$+ 848.8765\left(\frac{a}{t}\right)^5 - 876.9786\left(\frac{a}{t}\right)^6 + 370.8612\left(\frac{a}{t}\right)^7$$
(5.5)

The weight function for a single edge crack originating from the weld toe on a T-joint can be now determined directly in form of Eq. (5.2). This solution has been widely applied to welded T-plate joints [41] and cruciform joints.

Most of the related studies have used FEA and experimental tests to find residual stress distributions. Then, they calculated the SIF due to the residual stress (K_{res}) from FEM to overcome the use of the weight function. The final crack will be extended more than half plate thickness. That is longer than the restricted limits of relative depth (a/t) between (0.1-0.5), as mentioned in Refs. [41, 53] and (a/t) is between (0.1-0.7), as mentioned in Ref. [60].

Therefore, this work aims at showing the ability to use the weight function Eq. (5.2) with SINTAP profiles. Two approaches have been used to estimate K_{res}. In the first approach, the weight function is used to evaluate its ability for a/t greater than range (0-0.8), while in the second approach the residual stress distribution may be considered as a series of point loads along the expected crack line. Then, the residual stress intensity factors K_{res} are calculated based on fracture mechanics encoded using FRANC2D for each crack length, using the suitable boundary condition and point loads along the crack line.

5.5. Residual Stresses Distribution

Welding processes produce tensile residual stresses in components which have a significant effect on the integrity of structures. It is therefore important to have a detailed knowledge about the residual stress distributions in components and their redistribution after a crack has initiated in the residual stress field.

The existent defect assessment procedures, e.g., BSI 7910, SINTAP and R6, contain compendia with conservative estimation of the residual stress fields in a number of structures and materials. Further validation of the recommended residual stress profiles is, however, required.

A redistribution of stresses in the direction of the thickness gradually occurs as the crack is propagating. To assess the influence of the residual stresses on the failure of a welded joint, their distribution must be known. Hence, the failure can be characterized in terms of LEFM concept.

Although the plate butt welded geometry is the most studied in the literature, there are also limited informations on residual stress distributions in T-butt joints [69]. T-butt joints have the same stress distribution which can be used for cruciform joints; therefore, these joints (T-butt) are considered in this chapter.

The residual stress distributions recommended in the SINTAP compendium of residual stress profiles have different shapes for different weld geometries and orientations, including fourth and sixth order polynomials, cosine functions, and piece-wise linear distributions including bilinear and trapezoidal shapes.

In this work, four types of stress distributions have been examined; namely the BSI 7910 polynomial distribution (SINTAP), an approximate mean stress distribution which refers to least square linear fits, the proposed upper bound and lower bound distribution which refers to mean±2$STDV$, where $STDV$ stands for standard deviation [61]. Figure 5.2 shows the measurement of the transverse residual stresses of medium-strength and high-strength steel T-welded plate, which is widely used [60] for a plate thickness of 50 mm. The upper bound, lower bound and mean least square linear fit stress distributions are shown also in Figure 5.2.

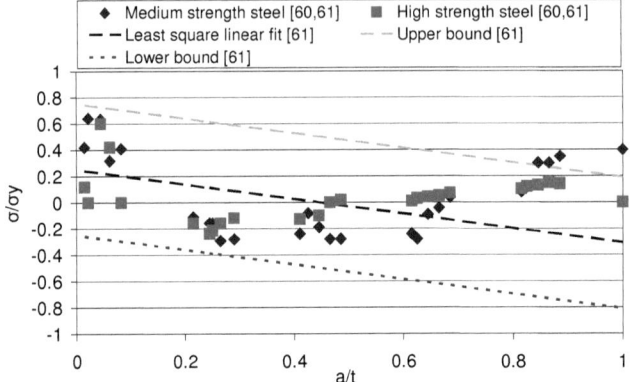

Figure 5.2: Transverse residual stress distributions for medium and high-strength steel T-plate, t=50 mm [60, 61]

In order to make a direct comparison between the different measured distributions, the distance is normalized by plate width t, and then the stress distributions are similar to each other, even if different thicknesses are used [60]. It may be seen that when stresses are normalized by yield strength σ_y the peak stresses are similar for different materials as shown in Figure 5.2, though for the high-strength plate, the normalized stresses are significantly lower at the weld toe [60]. Hyong et al. [61] used least square linear fits for different data sets that had been calculated for each of the T-plate and tubular joints.

The best fits of mean values, upper and lower bound lines at two standard deviations (*2STDV*) are shown in Figure 5.2, where *STDV* is equal to 0.25. The proposed normalised linear mean line for different joint data is given as [61]:

$$\frac{\sigma}{\sigma_y} = -0.56\left(\frac{a}{t}\right) + 0.25 \tag{5.7}$$

In general the crack length of interest in actual failure assessment is relatively short and well below $a/t = 0.5$ of the component thickness. Therefore, the crack lengths up to half thickness have been considered in Ref. [61], allowing the need to apply the stresses on the crack surface to a depth of $a/t = 0.5$ using the superposition rule, since the stress distributions over the region of $a/t > 0.5$ do not influence the SIF values. It is also worthy noting that the stresses in the region $a/t > 0.5$ are usually compressive and will have little effect on SIF values for short cracks [61].

5.6. BSI 7910 Distribution

There are two distributions for residual stress assessments in welded joints, namely R6 and BSI 7910 which are used in SINTAP [69] depending on the available informations about the welding conditions. If the weld informations are unavailable, then the BSI 7910 polynomial distribution is used. Otherwise, when welding conditions are known or can be estimated then the stress profiles given are based on the size of the plastic zone which correlates with the plate thickness, more details are found in Ref. [69]. In the current work, the BSI 7910 polynomial distribution was used. This distribution is given as:

$$\sigma/\sigma_y = \left(0.97 + 2.3267\,(a/t) - 24.125\,(a/t)^2 + 42.485\,(a/t)^3 - 21.087\,(a/t)^4\right) \quad (5.8)$$

Figure 5.3 shows the SINTAP distribution in value of (σ/σ_y=97%). The non-conservative values can be seen as compared with experimental measurements. SINTAP distribution (σ/σ_y =50%) has lower values and more satisfy with experimental data (see Eq. 5.9). In the current work the upper bound stress distributions (SINTAP=0.97, i.e. Eq. 5.8) has been used.

$$\sigma/\sigma_y = \left(0.5 + 2.3267\,(a/t) - 24.125\,(a/t)^2 + 42.485\,(a/t)^3 - 21.087\,(a/t)^4\right) \quad (5.9)$$

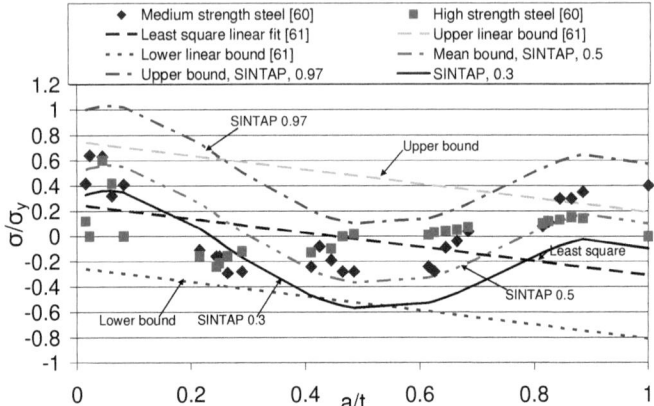

Figure 5.3: Transverse residual stress distributions compared with measured distributions of two steels, t=50 mm [60, 61]

Now if the residual stress distribution $\sigma(x)$ along the crack line is known, the corresponding SIF can be found by the weight function method (Eq. 5.1). In this work, the upper bound residual stress distributions are used.

Although the measured distributions are more precise, SINTAP profiles are recommended when no experimental measurements are available.

Therefore, it is assumed that the shape of the initial residual stress distribution corresponds to the SINTAP distribution (0.97) described in equation (5.8) can be adopted for conservative results.

5.7. Total Stress Intensity Factor

When the fatigue crack is propagating through a residual stress field in a welded plate, the stress intensity at the crack front is influenced by the combined effect of the local residual stress and the contribution resulting from the externally applied stress. The applied SIF is influenced by the weld geometry and the weld crack-like inclusions and is calculated from the fracture code FRANC2D as described thoroughly before.

Therefore, at crack tip, it appears that the actual SIF acting on a stress cycle (K_T) does not have the same magnitude as the applied SIF (K_{app}). The differences between the actual and the effective ΔK vary among different R-ratios.

In order to account the mean stress effect on crack growth rate da/dN, a method to estimate the effective level of an applied K for a given R-ratio has to be developed. It seems that the sound approach would be to develop a relationship between effective ΔK and R based on the concept of crack closure [96]. Different approaches have been developed to be sensitive with R ratio like NASGRO, Forman, Walker, etc.

The quantities assessment of the influence of a residual stress field on fatigue reliability can be determined by the principle of superposition based on the total SIF (K_T) for the solution of SIF due to the residual stresses and the solution of the SIF due to external loading such that:

$$K_T = K_{app} + K_{res} \tag{5.10}$$

The residual stress ratio, R_{res} in the situation under consideration is defined as:

$$R_{res} = \frac{\left(K_{min.app} + K_{res}\right)}{\left(K_{max.app} + K_{res}\right)} \tag{5.11}$$

$$\Delta K_{eff} = K_{max}(1 - R) \tag{5.12}$$

where $K_{min,app}$ and $K_{max,app}$ are the SIF due to minimum and maximum applied load, respectively. K_{res} is the SIF due to the residual stresses which is calculated according to the current approach. It should be noticed that the range of total stress intensity factor ΔK_T does not change when the residual stresses are superimposed such as [109, 110]:

$$\begin{aligned}\Delta K_T &= (K_{max})_T - (K_{min})_T = \left(K_{max,app} + K_{res}\right) - \left(K_{min,app} + K_{res}\right) \\ &= K_{max,app} - K_{min,app} = \Delta K_{app}\end{aligned} \tag{5.13}$$

and only the stress ratio, R is affected where $R_{app} \neq R_{res}$.

This indicates that the approach for crack growth in residual stress field should include the residual stress ratio R_{res}, in addition to the stress range $\Delta\sigma$, and SIF range ΔK. This approach requires the residual stress distribution and the associated stress intensity factor (SIF) which also depends on the fatigue crack length. It has been known that the welding residual stresses are distributed in a considerably non-uniform manner and are redistributed during the process of cyclic loading and fatigue crack growth.

The use of superposition has been criticized by some researchers because it considers only the initial residual stress field that exists in the un-cracked structures, with no acknowledge of the redistribution of residual stresses that occurs during the propagation of the fatigue crack. Therefore, Wu [28] mentioned that it is difficult to apply this approach for effective SIF evaluation because of the difficulty in capturing the actual residual stresses during crack propagation.

Other researchers have argued that the redistribution of residual stresses is of no consequence, e.g. LaRue et al. [110] showed that the slotting processes for purpose of starting a fatigue crack from a hole introduces a redistribution of the residual stresses although this did not have a considerable effect on the subsequent fatigue life.

Nevertheless to criticizing, superposition based technique is used extensively and it has been evaluated in the current work.

5.8. Crack Growth Propagation Life

The entire crack growth over the three regions needs to be mathematically approached to determine the fatigue crack growth like Forman, Walker, and NASGRO equations in terms of the range of total SIF. The latter is incorporated to calculate R_{res} using the superposition method.

5.9. The FNK Equation

The presence of residual stresses influences the crack growth rate da/dN value by considering the residual stress intensity ratio R_{res}. The FNK equation is considered the most suitable for the present study. Here, "FNK" stands for Forman, Newman and de Konig [96].

The equation is used in the most recent release of the crack growth prediction program, NASGRO. The NASGRO equation is written as:

$$\frac{da}{dN} = C_{ESA} \left[\left(\frac{1-f}{1-R} \right) \Delta K \right]^n \frac{\left(1 - \frac{\Delta K_{th}}{\Delta K}\right)^p}{\left(1 - \frac{K_{max}}{K_c}\right)^q} \tag{5.14}$$

where C_{ESA} and n are empirical parameters describing the linear region of the curve (similar to the Paris' model), and p and q are empirical constants describing the curvature in fatigue crack growth rate (FCG) data that occur near threshold (Region-I) and near instability (Region-III), respectively. The Newman's effective stress ratio (f), the threshold value of SIF range for a given R, (ΔK_{th}) and the critical SIF (K_c) are presented in Chapter Four above. The unit for the fatigue crack growth rate (FCG) da/dN is mm/cycle, and the SIF range ΔK is MPa (m)$^{1/2}$.

5.10. Results Verification and Procedures

A comprehensive program is established to calculate the fatigue life of welded joints using a language programming [111-115]. The input equation is SIF from FRANC2D. This approach includes:
1. Calculations of SIF due to the residual stresses.
2. Incorporations of different residual stress distributions.
3. Calculations of fatigue life and crack growth rate of welded joints, taking into account the effect of residual stress distributions, applied and residual stress ratio.

According to NASGRO equation, the crack propagation rate for a given R_{app} ratio was predicated by changing the R_{res}-ratio caused by varying K_{res} (Model-C). The latter was calculated as a product of the weight function and the residual stress distributions

(Eq. 5.1). The material data which are necessary for (Eq. 5.14) were obtained from AFGROW database [97] and Ref. [98], see Table 5.1 for steel S403.

The numerical integration has been carried out with help of FORTRAN language program to find K_{res}, and then the fatigue crack growth rate and the fatigue life are calculated. Two R ratios are incorporated in the program, namely, R_{res} and R_{app}. The former is calculated from Eq. (5.11) and incorporates the Newman's effective stress ratio which considers the effect of crack closure, ($f=K_{op}/K_{max}$). The latter, R_{app} is the nominal applied stress ratio (K_{min}/K_{max}).

The residual stress ratio (Eq. 5.11) gradually changes with the crack propagation when the residual stress distribution and K_{res} are considered in the evaluation of crack propagation. Therefore, in this work, K_{res} is calculated for each crack step together with the residual stress distribution using Eq. (5.1) for the vector x which is extended by step Δx inside each crack increment Δa (i.e., $x=0$ to a), $\Delta x << \Delta a$ and $x << a$. Then the superposition technique is used according to Eq. (5.10) to calculate the total SIF and to calculate R_{res}. The latter will be used in the fatigue life law (see Figure 5.4). It is obvious from Eqs. (5.10) and (5.14) that the fatigue life of the welded joint subjected to the effect of residual stresses and weld geometry can be evaluated if the solutions of SIFs, K_{res} and K_{app} are known.

With the current approach the da/dN-ΔK behavior of a welded component can be determined for a given base material data.

Very few investigations have dealt with NASGRO approach solution, with the effect of two parameters of stress ratios, namely applied and residual stress ratio.

Nguyen et al. [37] used Paris' law in their calculation which was based on the superposition of SIFs to include the effect of residual stresses on fatigue life calculations. Sobczyk et al. [109] used the most notable models that are the Forman equation in their calculations of crack growth rate and fatigue life.

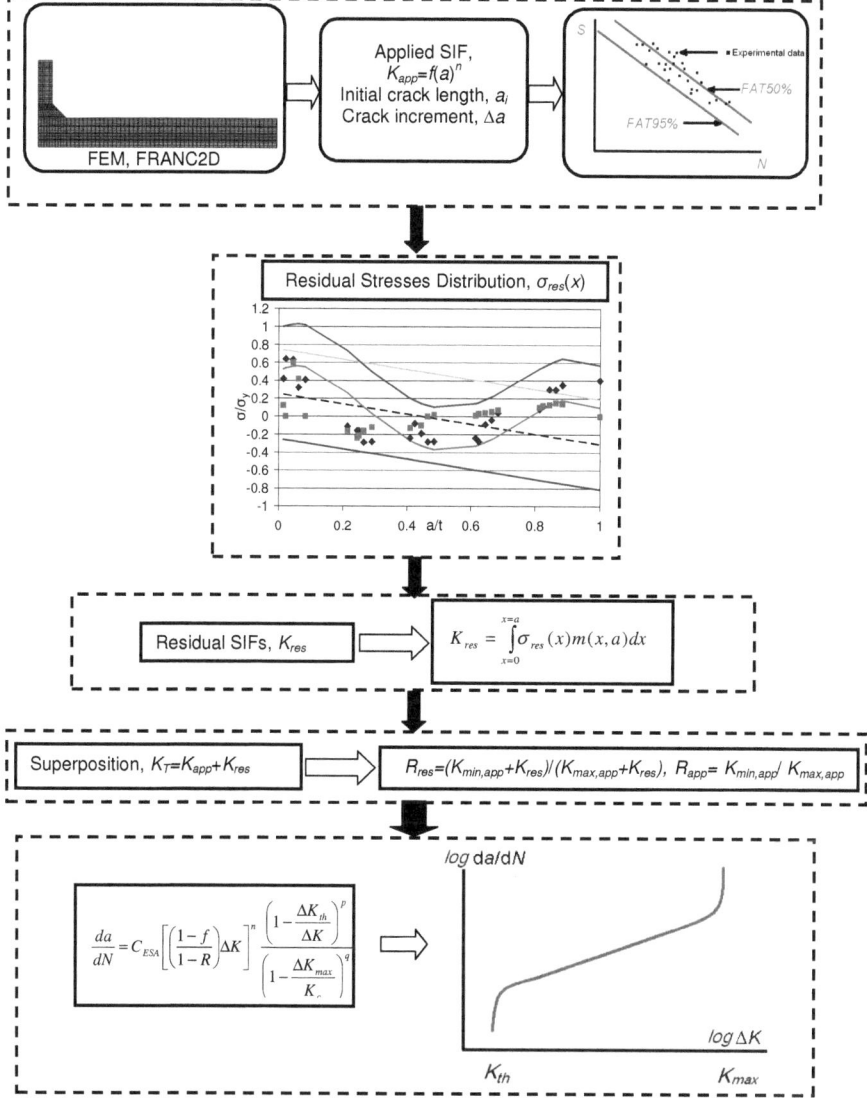

Figure 5.4: Procedures of fatigue life calculation under the effect of residual stresses

5.11. Effect of Residual Stresses Distribution on SIF

Several works have determined stress distributions using neutron diffraction data and FEM to calculate residual SIFs [41, 60, 61]. In the current work, the numerical integration is used to calculate K_{res} as a product of weight function and stress distribution. The current calculated SIF due to the residual stress (K_{res}) is compared with those obtained from Refs. [60] and [61].

The comparison between the numerical integration of Eq. (5.1) and FEM solution that is presented in the literature is shown in Figure 5.5 as based on 0.97, SINTAP distributions. It can be seen that the current solution of Eq. (5.1) agrees well with BSI 7910. K_{res} is calculated using a weight function solution and residual stress distribution that is given in the SINTAP procedure (BSI distribution).

The normalised residual stress intensity factor, K_{nor}, i.e., the SIF corresponding to the standard residual stress profile divided by $\sigma_y\,(t)^{1/2}$, is plotted versus relative crack depth (a/t) for the toe cracks at T-welded joints.

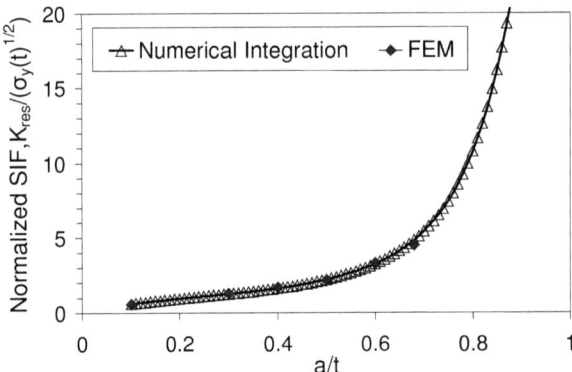

Figure 5.5: Numerical integration solution of residual SIF compared with FE results reported in [60]. The two solutions are based on BSI7910 distributions, SINTAP 0.97, $t=50$ mm, $a_i=5$ mm, $a_f=50$ mm, medium strength steel, $\sigma_y=345$ MPa

The different residual stress profiles of SINTAP are used in the current approach as shown in Figure 5.6, where still the BSI distribution has a conservative value. The solution of least square linear fits (Eq. 5.7) is more consistent with experimental data reported in Ref. [60].

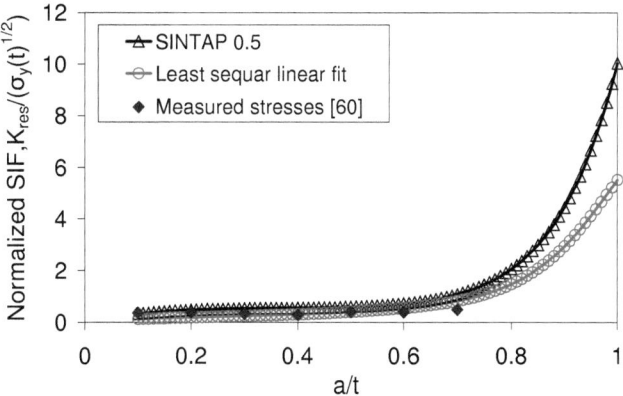

Figure 5.6: Numerical integration solution of residual SIF compared with that calculated as based on the measured stress distribution for medium strength steel [60], $t=50$ mm, $a_i=5$ mm, $a_f=50$ mm, medium strength steel, $\sigma_y=345$ MPa

It is shown that the solutions of the current procedure as based on the weight function and stress distributions agree well with FEA and experimental results that have been reported in the literature (see Figure 5.6).

The stress intensity factors due to the residual stresses have been calculated using Eq. (5.1) and FE. In the current work, the ability of both solutions are shown for a/t greater than 0.8.

The FE solution is less conservative and consistent with experimental data. The accuracy of the current analysis has been verified with experimentally measured data. The comparison of the K_{res} for T-plate welded joint calculated from experiments, FE, and by the current solution is shown in Figure 5.7. The SIF due to residual stresses, K_{res} increases as the crack length increases.

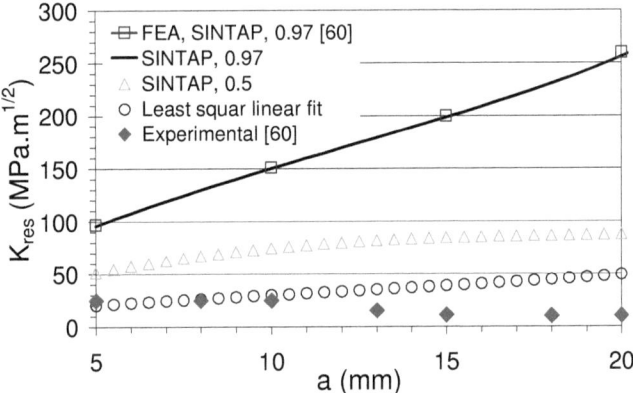

Figure 5.7: Calculations of K_{res} for different residual stress distributions of high strength steel, $a_i=5$ mm, $t=50$ mm, $\sigma_y=700$ MPa

O'Dowd et al. [60] used, in addition to above distributions, the R6 distribution which is given by a bi-linear distribution, when the heat input energy is known. Therefore R6 leads to non-conservative assessment. In this work, the SINTAP 0.5, and least square linear fit (Eq. 5.7) distributions are seen to be non-conservative

The calculated residual and applied SIF are superimposed to calculate the fatigue life. The calculations of applied SIFs have been described thoroughly using FRANC2D in the previous Chapter 3.

5.12. Predicting Fatigue Crack Growth Rates

Under cyclic loading only the stress ratio changes due to the presence of residual stresses. According to calculated K_{res} and ΔK_{app} the new R ratio can be assumed is called as residual stress ratio, R_{res}.

Considering the welding residual stress effects in accordance to the crack closure phenomenon the NASGRO equation is applied (Eq. 5.14). Table 5.1 lists the material constants data for structural steel, and welding joints with $\alpha=2.5$, $S_{max}/\sigma_o=0.3$. These constants are needed in fatigue life prediction within NASGRO equation. In case of complete penetration, the weld toe cracks ($a_i=0.1$ mm) are assumed to start and propagate through the thicknesses.

In fact, the initial crack length has to be unified as calculated in the previous Chapter Four. The weld toe crack has a length of 0.1 mm and the root crack has the length equal to LOP.

Table 5.1: Constants for structural steel, shipbuilding steel and the welding joints [98]

Steel	ΔK_o	K_c	C_{ESA}	n	p	q	$C_{th}+$	$C_{th}-$
S235	6	45	10^{-8}	3	0.5	0.5	1.9	0.1
S460	6.5	70	10^{-8}	3	0.5	0.5	1.9	0.1
S690	5.1	98	5×10^{-8}	2.3	0.5	0.5	1.9	0.1
S325	8.7	40	5×10^{-8}	3.3	0.25	0.25	3	0.25
S283	9	33	4×10^{-8}	3.3	0.25	0.25	2	0.25
S403*	8.7	30	4×10^{-8}	2.2	0.25	0.25	2.7	0.25
S885 MB	5.11	106	2.4×10^{-8}	2.7	0.25	0.25	1.9	0.1
S885 HAZ	7.68	98	2×10^{-8}	2.5	0.5	0.5	1	0.1
S885WM	5.6	70	4×10^{-8}	2.5	0.5	0.5	2.5	0.1
S960MB	5	60	5×10^{-8}	2.5	0.5	0.5	1.9	0.1
S960HAZ	7.8	75	4.5×10^{-8}	3.1	0.6	0.25	1.9	0.1
S960WM	5.5	62	1.3×10^{-8}	2.8	0.5	0.5	1.9	0.1

*The used steel is S403.

Figure 5.8 shows the predicted fatigue crack growth (FCG) rate for nominal or applied stress ratios ($\sigma_{min}/\sigma_{max}$) of $R= 0.1$, $R= 0.3$ and $R=0.5$ at a constant applied stress range of 200 MPa. The material used in crack growth rate prediction is steel S403. The unit for the fatigue crack growth rate da/dN is mm/cycle, and the SIF range ΔK is MPa $(m)^{1/2}$.

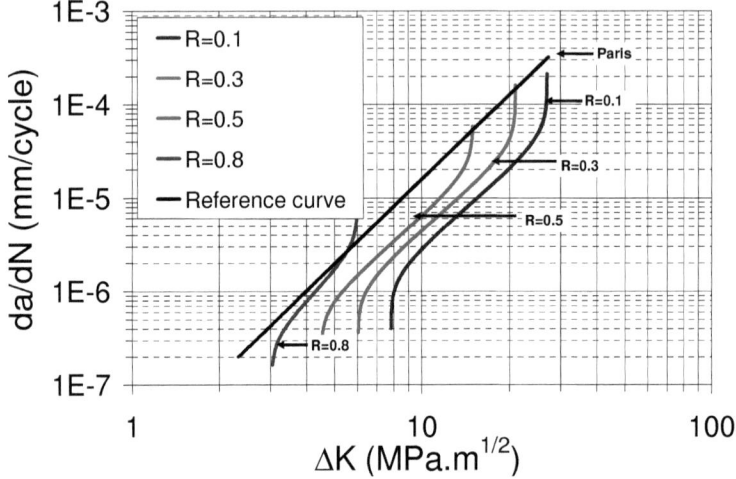

Figure 5.8: Comparison between predicted FCG rates for three R ratios and Paris' solution for steel S403 without residual stresses, $a_i=0.1$ (toe crack), $\Delta\sigma=70$ MPa. Reference curve from Paris' solution with $C_{char}=5E-13$, $m=3$

NASGRO has also taken into account also the final fast crack growth stage when K_{max} approaches K_c. Also, a good prediction has been given using NASGRO for different R ratios, due to the number of material fitting constants which give the ability to present the three regions of crack growth rate. Due to nature of NASGRO equation, the three regions of crack growth rate are shown in Figure 5.8.

The IIW presents the conservative data of crack growth rates owing to test at high values of R ratios as mentioned above. Figure 5.9 shows the comparison between the crack growth curve using Paris' law ($C=5E-13$ and $m=3$ from IIW) and the curve from NASGRO equation (C_{ESA}, n and fit parameter's from Table 5.1) by applying a high R ratio. It is shown that Paris' curve provides a conservative analysis.

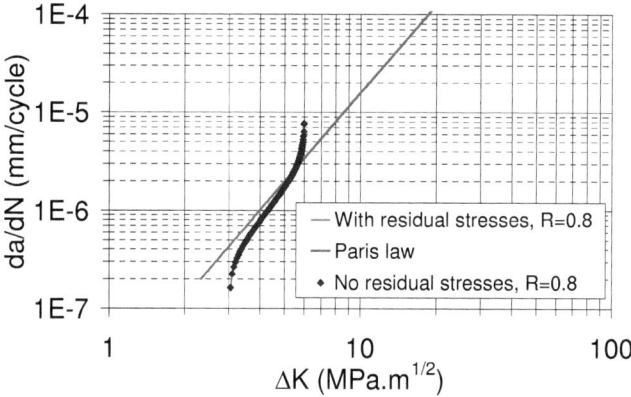

Figure 5.9: Comparison between two approaches at $\Delta\sigma=70$ MPa, $a_i=0.1$ mm, $t=12$ mm. Paris' constant are $C_{char}=5E-13$ and $m=3$ [3, 34], S403

It is to be emphasized that the tendency towards an higher crack growth rate in region-III arises from the influence of static crack extension mechanisms such as micro-cleavage and void coalescence, where the final failure mechanism is one of cleavage fracture, contributions to crack growth by micro-cleavage generally occur when K_{max} approaches K_{mat} (typically $K_{max} > 0.7\ K_{mat}$) [107].

5.13. Effect of Residual Stresses and Stress Ratio on FCG Rate

The effect of residual stresses on fatigue crack growth rate (FCG) is accounted by replacing the nominal R ratio with residual stress ratio R_{res}. The latter is calculated using the superposition of applied SIF (K_{app}) and SIF due to residual stresses, K_{res}. NASGRO considers the effect of plasticity induced crack closure by an empirical constant f as a function of R_{res}. In fact, the crack closure phenomenon incorporates the effect of residual stresses. Therefore, the new R_{res} will be used within the crack closure ratio, f. More details have been presented in Chapter Four.

Figure 5.10 shows that the effect of residual stresses on the crack growth rate is negligible as the applied R ratio increases. This is because the tensile residual stress reduces the effect of crack closure and increases the crack growth rate. Therefore, for higher applied R ratios, the influence of residual stresses on growth rate is small because the crack remains open throughout the whole cycle.

The effect of stress ratio can be discussed by the effect of crack closure mechanism. This mechanism is related to the crack path meanders. It is important to explain stress ratio effects in connection with residual stresses as shown in Figure 5.10 in terms of mode-I stress intensity factor range ΔK_I. At higher R-values, the curves of da/dN (Figure 5.10a-e) differ from each other substantially in the threshold region, showing a typical stress ratio effect.

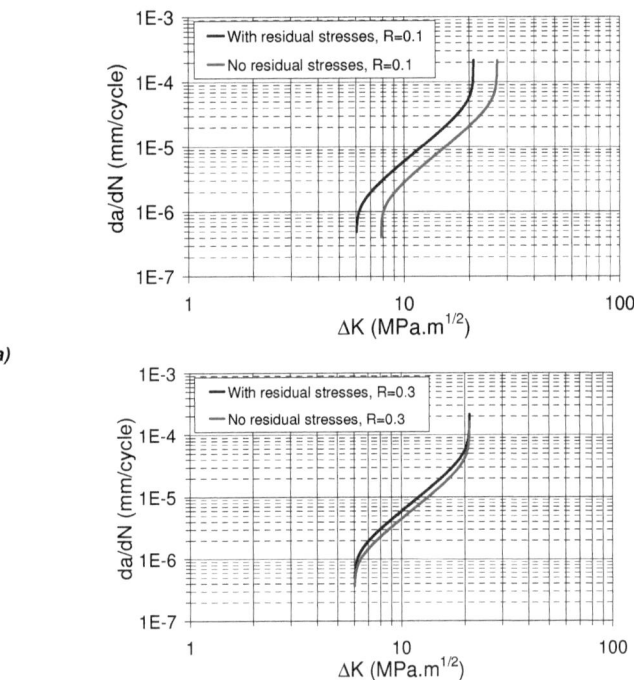

(a)

(b)

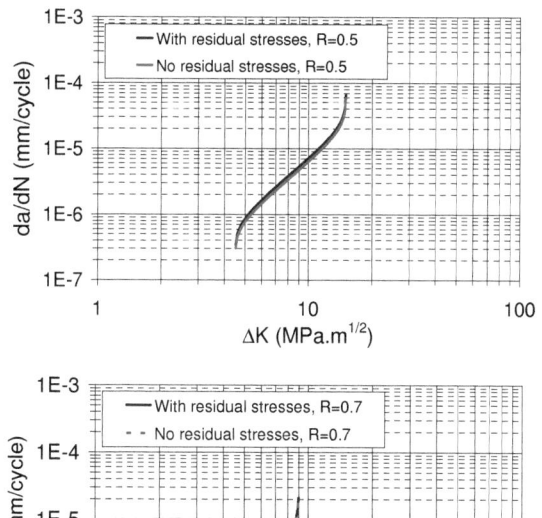

(c)

(d)

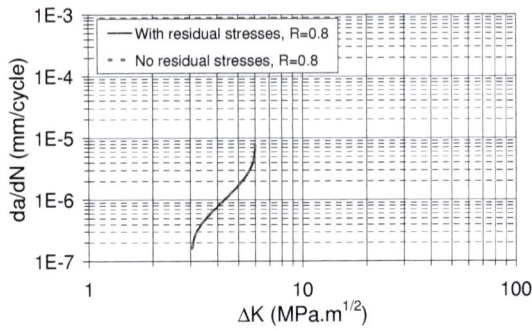

(e)

Figure 5.10: *Effect of applied R ratio and existence of residual stresses on the predicted FCG rate, $t=12$ mm, $a_i=0.1$ mm, $\Delta\sigma=70$ MPa, $\sigma/\sigma_y=0.97$, steel S403: (a) $R=0.1$; (b) $R=0.3$; (c) $R=0.5$; (d) $R=0.7$; (e) $R=0.8$*

In literature, the residual stresses are usually assumed negligible in small thickness scale or relieved [31, 84]. In small scale specimens, residual stresses are small [3, 34]. According to IIW, the results can be corrected to allow for the greater effects of residual stresses in real components and structures. This may be achieved either by testing at high R ratios e.g., $R=0.5$ or by testing at $R=0$ and lowering the fatigue strength at 2 million cycles (FAT) by 20% [3, 34]. In this work, R ratios between 0.5-0.8 have been estimated in which the effect of residual stresses is negligible as shown in Figure 5.10(c-e). These R values can be used in testing.

For more details, the residual stress profiles for different geometries are presented in Ref. [69]. The compendium (from SINTAP) obtained from the literature review is valid for some ranges of thickness and geometry. The butt and T- and/or cruciform joints are the most used in engineering structures. The butt and T-joint or cruciform joints with thickness range in mm (24-300) and (25-100) are presented respectively in the residual stresses compendium [69].

Figure 5.11 shows the predicted fatigue crack growth (FCG) rates for the case $R=0.1$ at a constant applied stress range of 70 MPa. The effect of residual stresses has been taken into account for thickness 26 mm. Hence the crack growth rate will increase with existing residual stresses. The crack length as a function of number of cycles from NASGRO is shown in Figure 5.12.

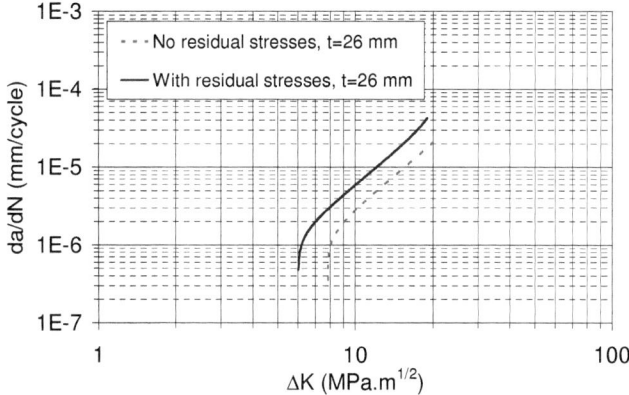

Figure 5.11: *Effect of sheet thickness for steel S403 at $\Delta\sigma=70$ MPa, $R=0.1$, $a_i=0.1$ mm, $t=26$ mm*

There is no difference between the two stress distributions (97% and 30%) in the case of residual stresses being taken into account, see Figure 5.12.

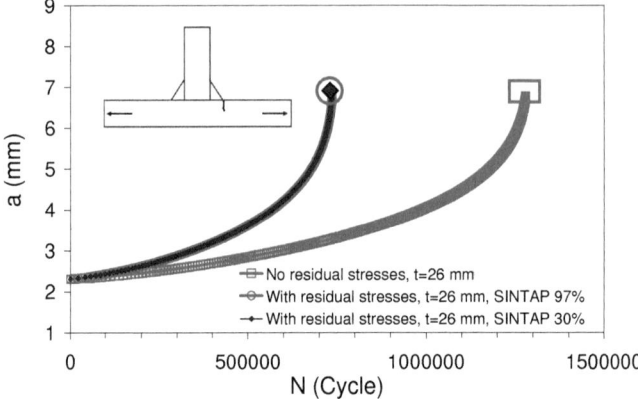

Figure 5.12: *Fatigue crack growth curves, steel S403 at $\Delta\sigma=70$ MPa, $R=0.1$, $a_i=0.1$ mm. Residual stress distribution $\sigma/\sigma_y = 0.97$ (SINTAP 97%) and post weld heat treatment stress distribution $\sigma/\sigma_y = 0.3$ (SINTAP 30%)*

According to Ref. [31] there is no significant difference in the thickness effect between AW (as welded) and PWHT (post weld heat treated) welds. In turn, the different residual stress distributions have been used for AW and PWHT (as shown in Figure 5.12).

It is concluded that the effect of residual stresses in terms of crack closure will be negligible in the case of high stress ratios, as shown in Figures 5.13 and 5.14.

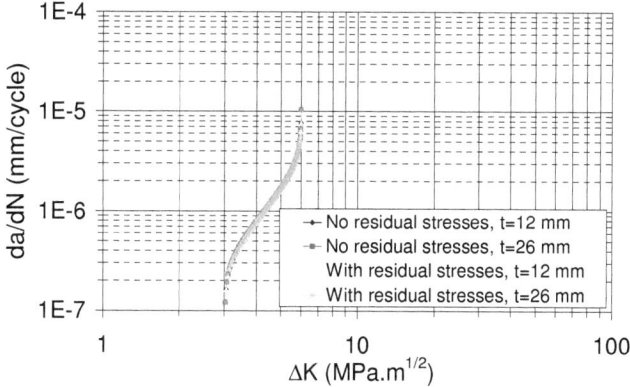

Figure 5.13: *Effect of plate thickness for steel S403 steel at $\Delta\sigma=70$ MPa, $R=0.8$, $a_i=0.1$ mm*

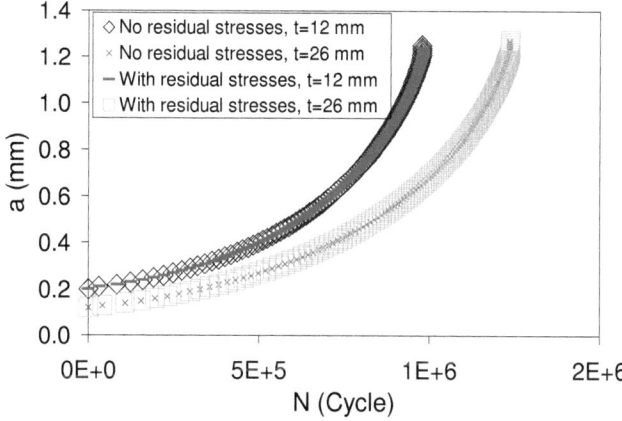

Figure 5.14: *Fatigue crack growth curves for steel S403 steel at $\Delta\sigma=70$ MPa, $R=0.8$, $a_i=0.1$ mm, SINTAP 97%*

Nykänen et al. [15] predicted FAT values for different welded joints. They expressed the fatigue behaviour in terms of stress range where the crack assumed to be under residual tensile mode. These residual stresses keep the crack tip open during the stress cycle. No residual stress distributions were assumed. To evaluate Nykänen's work using the current model and to study the effect of residual stresses, transverse butt weld joints with partial penetration have been used with LOP=5.5 mm, where LOP/t=0.22 as described in Chapter Four (see Figure 4.13).

Figure 5.15 shows the a-N curves for a transverse butt weld joint from Nykänen [15]. The Model-C refers to use NASGRO equation. Two possibilities can be studied with this model, either to include residual stresses (upper bound) or without residual stresses. Model-A refers to use the Paris' law with R=0.5, C_{char}=5E-13, and m=3 as recommended by IIW [3, 34]. The transverse residual stress profiles of transverse butt weld joints are given in Ref. [69]. The relevant equation is:

$$\sigma/\sigma_y = \left(1 - 0.917(a/t) - 14.533(a/t)^2 + 83.115(a/t)^3 - 21545(a/t)^4 + 24416(a/t)^5 - 9636(a/t)^6\right) \quad (5.15)$$

The distribution of residual stresses (SINTAP 30%) is also included and shown in Figure 5.15. The same results are obtained as compared with the case of toe crack in cruciform joints (see Figure 5.12). There are no significant differences between as welded residual stress distributions (upper bound) and the post weld heat treatment residual stress distribution (30%) in case of root crack of transverse butt weld joint. It is to be emphasized that under tension-tension cyclic load conditions, residual stresses did not show a significant effect on the fatigue life of welded joint.

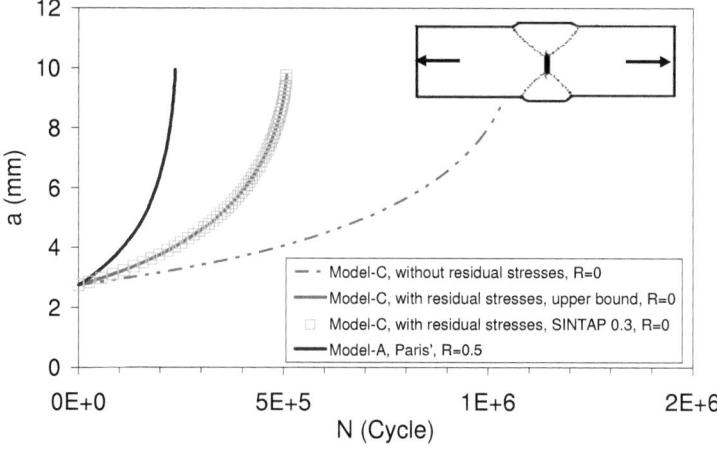

Figure 5.15: Fatigue crack growth curves of transverse butt weld (see Figure 4.12) at $\Delta\sigma=100$ MPa, $LOP=2a_i=5.5$ mm, $LOP/t=0.22$, thickness=25 mm

5.14. S-N Curve

The materials constants used are those for the steel S403, Table 5.1. Figure 5.16 shows the *S-N* curve for different sheet thicknesses using NASGRO equation as compared with the solution from Paris' law. As described before, the initial crack length has been used in case of weld toe crack equal to 0.1 mm.

The effect of sheet thickness and residual stresses under high stress ratio can be considered as negligible. However, there are small differences between them probably due to numerical integration routines, as shown in Figure 5.16.

The higher fatigue life curve of Paris' law is calculated for *C* and *m* from IIW [3, 34]. Therefore, there are some differences between this solution and the solution from NASGRO.

Paris empirical constants (*C* and *m*) are developed at an $R=0.8$ as an upper bound to the FCG curves established from NASGRO (see Figure 5.9).

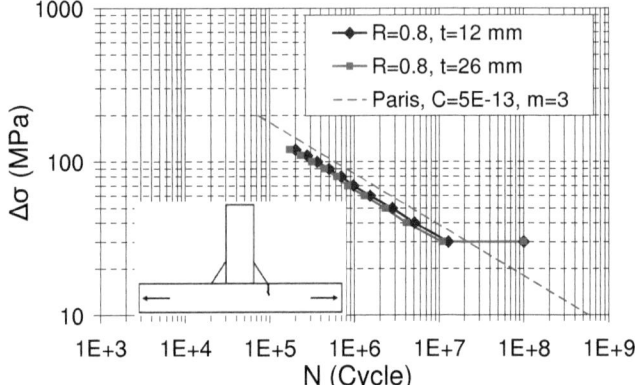

Figure 5.16: Comparison between NASGRO and Paris' fatigue life solutions without residual stresses

At high stress ratio ($R=0.8$), the effect of residual stresses are also shown to have a small effect on the NASGRO solutions of the two thicknesses, however the endurance limit has been decreased for the thicker plate (see Figure 5.17). This observation is shown under the effect of residual stresses, where definitely the thicker plate will have higher residual stresses and hence lower life as shown in Figure 5.17.

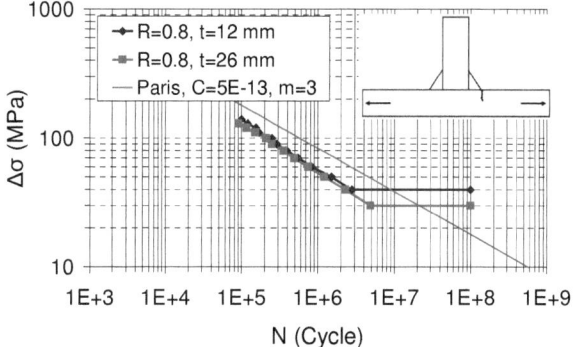

Figure 5.17: Comparison between NASGRO and Paris' fatigue life solutions with assumption the presence of residual stresses

At a low stress ratio of $R=0.1$, the effects of residual stresses are obvious. Therefore, the thicker joints with the consideration of residual stresses will have a lower fatigue life as shown in Figure 5.18.

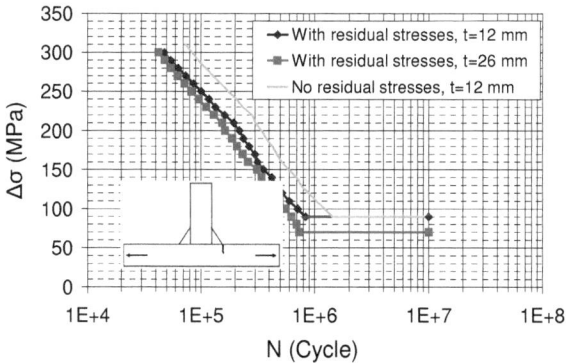

Figure 5.18: S-N curves at $R=0.1$, $a_i=0.1$ mm, SINTAP 97%

It is evident from Figures 5.17 and 5.18 that the endurance stress range ($\Delta\sigma_e$) for a smaller plate thickness (12 mm) is higher than that for plate thickness (26 mm) due to the existence of residual stresses.

5.15. Conclusions

Most studied have neglected the effect of residual stresses in small thickness scale. In this work a numerical integration program is established to calculate SIFs due to residual stresses (K_{res}) which is a product of traditional weight function and residual stress distributions through the thickness of the plate. The transverse residual stresses have a major effect on the crack growth through the thickness of joints. Two transverse residuals stress distributions away from the weld toe and from LOP are used for the plate T-butt weld and for plate butt welds. For conservative results, the upper bound residual distributions are used in this work. For evolution purpose, the other distributions are examined.

The residual stress distributions which are given in the recommendations are used in the current procedures and compared with literature. Due to limiting use of weight function, most literature has used the residual stress distributions from experimental measurements (neutron diffraction) and FEM to calculate K_{res}. In this work, the calculations which are based on the weight function and available distributions to calculate K_{res} and lives are satisfactory.

Since the weight function is used for a pre-existing crack, the application of this function method to the crack propagation analysis step by step with the presence of a residual stress field is yet to be carried out.

The solution of SIF due to the residual stress (K_{res}) from the product of weight function and SINTAP residual stress distributions have been compared with those obtained from FEM-based methods and measured experimental data. The comparisons show a good agreement. Therefore, the ability to use the weight function higher than the restricted limits which were mentioned before ($a/t=0.5$) is demonstrated.

The phenomenon of the change of the residual stresses during fatigue cycle is investigated in correlation with the applied load ratio using the superposition method. In this work, the superposition of residual stresses in the calculation of total SIF is used. Although this method has been criticized by other researchers, the results which were obtained are quite well.

An increasing R ratio will reduce the effect of residual stresses on the predicted life because the higher tensile residual stresses reduce the effect of crack closure, i.e., the crack remains open. Therefore, for higher R ratios, the influence of residual stresses and K_{res} on the growth rate is small. Important stress ratio effects appear in the threshold region, where ΔK_{th} decreases as R increases.

The comparisons with literature in case of residual stresses are presented in this chapter. Different distribution ratios were used for this comparison, e.g. σ/σ_y=0.97%, 0.5% and 0.3%.

Finally, the effects of joints geometry and materials properties can be studied according to the current approach.

Chapter Six
CONCLUSIONS AND RECOMMENDATIONS FOR FUTURE WORK

6.1. Conclusions

This work has three main parts: in the first part, the stress intensity factor (SIF) is calculated and evaluated; in the second part the SIF is used in the calculations of fatigue life; and by using suitable life formula, backward calculations using the built language program were carried out to determine the critical crack sizes (a_i, and a_f), and fatigue strength (FAT). The fatigue life calculations with the effect of residual stresses are described in the third part.

Even though there are many studies dealing with fracture mechanics, this book combines FEA with fracture mechanics to develop a simple and reliable procedure to predict fatigue life of welded joints on basis of fatigue crack growth. The used approach works within the assumptions of linear elastic fracture mechanics (LEFM). The adopted method assumes a crack existing at an edge of welded joints. After this crack starts to grow, it will be under the normal applied tensile stress (maximum stress criterion). Hence, the effect of plasticity will be small when the opening mode dominates. Therefore, the assumption of LEFM is reasonable.

Fatigue life calculations of welded joints with different combinations of geometrical parameters have been carried out using the results of SIFs that were obtained from FEM. These SIF results are evaluated, benchmarked and compared with available solutions. New SIF solutions have been developed then.

Chapter Six   *Conclusions and Recommendations for Future Work*

The simulation of fatigue crack propagation and SIF calculations using FEM is found to be effective to determine the fatigue strength (FAT) and establishing the *S-N* curve. Experimental data from literature are used to verify the current approach including the effects of weld geometries and residual stresses.

The fatigue life and SIFs of load-carrying joints with various weld shapes have been accurately estimated by performing a fatigue crack propagation analysis when the crack is originated from the weld root of the joint or from the lack of penetration (LOP), respectively. In contrast, the fatigue life of non-load-carrying joints is investigated with the presence of the weld toe crack.

Although real components are often different from the details contemplated in the standards, successful comparisons are carried out with experimental data. Moreover, this work investigates the effect of some welded geometrical parameters and residual stresses on the propagation life of welded joints. Verifications of FAT classes from standers are also carried out.

To carry out the integration of the fatigue life equation (e.g., Paris or NASGRO), accurate values of initial crack length, and final crack length with a fixed crack growth parameters (C and m) are required. The values of cracks length are determined and compared with experimental data. A value of the initial crack length of 0.1 mm is shown to be a typical length for the case of weld toe crack. Longer crack lengths are determined in the case of a root crack where the crack length is equal to the size of the non-welded area (LOP).

Although there are considerable fatigue data for welded joins for steel and aluminum in recommendations such as IIW and BSI, the problem exists to find the fatigue strength data for some other materials. In that case, the sparse design data for welded joints for different materials can be calculated easily with the current work.

The results are then used to indicate the initial weld sizes which would be necessary to avoid failure on the weld throat (case of LOP) and on the base plate (case of toe crack). Therefore, the entire fatigue life is assumed to be crack propagation.

As expected, increasing the size of the lack of penetration will increase the probability of weld throat crack propagation and will decrease the fatigue life absolutely. Also, the increasing weld size will decrease the FAT value.

Verification of the current results is reported during the comparisons with experimental data. Moreover, the comparisons with predicted FAT-class from Ref. [15] are presented.

The recommendation to use the current approach parameters will take an advantage, because SIFs and a_i are only needed to find an appropriate FAT for different geometries.

The crack propagation of two cracks at same time has different criteria for estimation the crack life, i.e., determination which crack will contribute to final failure, or if both will contribute in final failure. The faster crack will propagate according to throat thickness geometry and applied load nature.

Residual stresses have an effect on fatigue crack propagation. The residual stress distributions are studied and compared with the experimental data. Therefore, SIFs due to residual stresses are calculated from the product of weight function and residual stress distributions at the crack tip. The universal weight function was used in this analysis.

These calculations we carried out step by step for each cycle and each crack increment.

Residual stress ratios were calculated as a function of the residual SIF (K_{res}) that were calculated from the weight function method and applied SIF (K_{app}) that in turn were calculated from FEM. The combination of residual stress ratio and applied stress ratio with residual stress effects are shown to be less investigated and unexplained in the literature for fillet welded joints.

This work has shown this combination which gives in turn the crack propagation under the effect of residual stresses. The main finding is that the effect of residual stresses and sheet thickness are ignorable when the stress ratio is more than $R=0.5$. In the case of lower stress ratio ($R=0.1$), a lower sheet thickness will have a higher fatigue strength. These results also reveal that the residual stresses increase with increasing the component thickness due to the increase in the heat input of the welding in the real workshop, which in reality means that a larger thickness will have a temperature gradient that increases the residual stresses.

6.2. The Originality of the Current Calculations

Although, the application of ordinary fracture mechanics parameters like K to very small cracks is not appropriate (particularly in fatigue and short crack growth), this work presents good results according to LEFM. The original parts from this book are summarized by the following points:

1. Examination the crack path effects on SIF solution and in turn it's effect on fatigue life. Therefore, the difference between the SIF-solutions from IIW, BSI and current work has been presented and their effect on S-N is shown. Then the effect of meandering cracks is taken into account.

2. Determined analytically the crack's lengths which seriously effect joints strength and lead to catastrophic failure. Moreover, these lengths are evaluated as compared with experimental data. Typical crack lengths were determined according to the joint's geometry, crack's type and locations. Initial crack length equal to 0.1 mm and LOP are satisfied in case of toe and root crack respectively. Final crack length and crack increment also verified. In previous works, these lengths are presented mostly as a range of length or they have been assumed [15].

3. Determined the fatigue strength and life of unknown cases. The validations of FAT-values are examined for different notch joints. Also the effects of joints geometry such as plate thickness are investigated. The explanation of the thickness effect using the current work is presented according to fracture mechanics and FE point of view. Then, the effect of constant crack length and scaling crack length with respect to thickness is presented.

4. The effects of weld notch geometry were investigated as well as their effect of FAT-value.

5. Calculated analytically the SIF due to residual stresses which previously calculated either from experimental measurements of residual stress distributions, or from FEM.

6. The combination of two R-ratios (residual and applied stress ratio), and calculating SIF due to residual stresses for each crack step have not yet been sufficiently investigated. In addition, these calculations were repeating at crack tip for each crack increment.

7. An increasing R ratio will reduce the effect of residual stresses on the predicted life because the higher tensile residual stresses reduce the effect of crack closure, i.e., the crack remains open. Therefore, for higher R ratios, the influence of residual stresses and K_{res} on the growth rate is small. Important stress ratio effects appear in the threshold region, ΔK_{th} decrease as R increase.

8. The comparison with Nykänen's work shows that there is an effect of residual stresses distributions and stress ratio.

9. The stress concentration factor can be calculated from predicted FAT-values. Then the machined weld overfills give higher fatigue resistance.

6.3. Recommendations for Future Work

For time and experiments constraints and according to the lack found in the related literature, some works are recommended for future work such as:

1. No standards are presently available for welded components made of other materials (e.g. titanium alloys, etc.). The current fracture mechanics approach can be applied for other materials.

2. Investigation of other welded joints' geometries and calculation the FAT values for layered joints such as spot welded joints.

3. Simulation of the various types of defects within the weld using the current approach. These defects might have an effect on the calculated value of the crack length. Therefore, a new systematic set of values and relationships can be developed.

4. Studying the validity of the currently calculated parameters (C, m, a_i, a_f, FAT and S-N curve) with other loading types.

Chapter Six   **_Conclusions and Recommendations for Future Work_**

5. Studying the validity of the currently calculated parameters under the effect of environment, e.g., corrosion effect, temperature and cryogenic effect.

6. Establishing a criterion to forecast the possible crack initiation and propagation region (toe or root) in the welded joints which have these two types of cracks, i.e., determine the faster effecting crack causing the final failure. Also to determine the probable FAT and *S-N* curve.

REFERENCES

[1] Lassen T., and Recho N., Fatigue Life Analyses of Welded Structures, ISTE Ltd, 2006.

[2] Blondeau R., Metallurgy and Mechanics of Welding, ISTE Ltd, 2008.

[3] Hobbacher A., Recommendations for Fatigue Design of Welded Joints and Components, IIW Doc., No XIII-1965r14-03 / XV-1127r14-03, International Institute of Welding, 2006

[4] Branco C.M., Maddox S.J., Infante V. and Gomes E.C., Fatigue Performance of Tungsten Inert Gas (TIG) and Plasma Welds in Thin Sections, International Journal of Fatigue, Vol. 21, No. 6, pp. 587–601, 1999.

[5] Barsoum Z. and Barsoum I., Residual Stress Effects on Fatigue Life of Welded Structures Using LEFM, Engineering Failure Analysis, Vol. 16, No. 1, pp. 449–467, 2009.

[6] Nagaralu R., Fatigue Crack Growth under Residual Stress around Holes, Mississippi State University, M.Sc. Thesis, USA, 2005.

[7] Andersen M.R., Fatigue Crack Initiation and Growth in Ship Structures, Ph.D. Thesis, Technical University of Denmark, Denmark, 1998.

[8] Alam M.S., Structural Integrity and Fatigue Crack Propagation Life Assessment of Welded and Weld-Repaired Structures, Ph.D. Dissertation, Louisiana State University, USA, 2005.

[9] Webster G.A., and Ezeilo A.N., Residual Stress Distribution and Their Influence on Fatigue Life Times, International Journal of Fatigue, Vol. 23, No. 1, pp. s375–s383, 2001.

[10] Fukuda S. and Tsuruta Y., An Experimental Study of Redistribution of Welding Residual Stress with Fatigue Crack Extension, Transaction of JWRI, Vol. 7, No. 2, pp. 67–72, 1978.

[11] Lampeas G.N. and Diamantakos I.D., Calculation of Stress Intensity Factors of Cracked T-Joints Considering Laser Beam Welding Residual Stresses, First International Conference on Damage Tolerance of Aircraft Structures, TU Delft, The Netherlands, September 25–28, 2007.

[12] Eurocode 3, Design of Steel Structures, 1990.

[13] Germanischer Lloyd Aktiengesellschaft, Rules for Classification and Construction, Part I Ship Technology, Section 20C, Fatigue strength, Germanischer Lloyd Aktiengesellschaft, GL 2007, <www.gl-group.com>.

[14] British Standards Institution, Guidance on Methods for the Acceptance of Flaws in Structure, PD 6493, BS 7910, Appendix J, 2005.

[15] Nykänen T., X. Li, Björk T. and Marquis G., A Parametric Fracture Mechanics Study of Welded Joints with Toe Cracks, Engineering Fracture Mechanics, Vol. 72, No. 10, pp. 1580–1609, 2005.

[16] Smith I.F.C. and Smith R.A., Fatigue Crack Growth in a Fillet Welded Joints, Engineering Fracture Mechanics, Vol. 18, No. 4, pp. 861-869, 1983.

[17] Testin R.A., Yung J.Y., Lawrence F.V. and Rice R.C., Predicting the Fatigue Resistance of Steel Weldments, Weld. Res. Suppl., Vol. 4, pp. 93–98, 1987.

[18] Ferreira J.M. and Branco C.M., Influence of Fillet Weld Joint Geometry on Fatigue Crack Growth, Theoretical and Applied Fracture Mechanics, Vol. 15, pp. 131–142, 1991.

[19] Motarjemi A.K., Kokabi A.H., Ziaie A.A., Manteghi S. and Burdekin F.M., Comparison of the Stress Intensity Factor of T and Cruciform Welded Joints with Different Main and Attachment Plate Thickness, Engineering Fracture Mechanics, Vol. 65, No. 1, pp. 55–66, 2000.

[20] Balasubramanian V. and Guha B., Fatigue Life Predication of Welded Cruciform Joints Using Strain Energy Density Factor Approach, Theoretical and Applied Fracture Mechanics, Vol. 34, No. 1, pp. 85–92. 2000.

[21] Frank K.H. and Fisher J.W., Fatigue Strength of Fillet Welded Cruciform Joints, J. Struct. Division, ASCE, 105(ST9), pp. 1727–1740, 1979.

[22] Usami S. and Kusumoto S., Fatigue Strength at Roots of Cruciform, Tee and Lap Joints, Transactions of Japan Welding Society, Vol. 9, No. 1, pp. 1–10, 1979.

[23] Caccese V., Blomquist P.A., Berube K.A, Webber S.R. and Orozco N.J., Effect of Weld Geometric Profile on Fatigue Life Cruciform Welds Made by Laser/GMAW Processes, Marine Structure, Vol. 19, No. 1, pp. 1–22, 2006.

[24] Teng T.-L., Fung C.-P. and Chang P.-H., Effect of Weld Geometry and Residual Stress on Fatigue in Butt-Welded Joints, International Journal of Pressure Vessels and Piping, Vol. 79, No. 7, pp. 467–482, 2002.

[25] Mashiri F.R., Zhao X.L. and Grundy P., Effect of Weld Profile and Undercut on Fatigue Crack Propagation Life of Thin Walled Cruciform Joint, Thin Walled Structures, Vol. 39, No. 3, pp. 261–285, 2001.

[26] Maddox S.J., An Analysis of Fatigue Cracks in Fillet Welded Joints, International Journal of Fracture, Vol. 11, No. 2, pp. 221–243, 1975.

[27] Metrovich B., Fisher J.W., Yen B.T., Kaufmann E.J., Cheng X. and Ma Z., Fatigue Strength of Welded AL-6XN Superaustenitic Stainless Steel, International Journal of Fatigue, Vol. 25, No. 9, pp. 1309–1315, 2003.

[28] Wu W., Fatigue Crack Propagation Behavior of Welded and Weld Repaired 5083 Aluminum Alloy Joints, M.Sc. Thesis, School of Aerospace and Mechanical Engineering, The University of New South Wales, Australia, 2002.

[29] Poutiainen I. and Marquis G., A Fatigue Assessment Method Based on Weld Stress, International Journal of Fatigue, Vol. 28, No. 9, pp. 1037–1046, 2006.

[30] Engesvik K., Analysis of the Uncertainty of the Fatigue Capacity of Welded Joints, NTNU, Trondheim, UR-82-17, Norway, 1983.

[31] Lindqvist J., Fatigue Strengths Thickness Dependence in Welded Construction, M.Sc. Thesis, Borlänge, Sweden, 2002.

[32] Karlsson N. and Lenanader P.-H., Analysis of Fatigue Life in Two Welded Class System, M.Sc. Thesis, LITH-IKP-EX—05/2302-SE, Department of Mechanical Engineering, Linköping University, Sweden, 2005.

[33] Hobbacher A., Stress Intensity Factors of Welded Joints, Engineering Fracture Mechanics, Vol. 46, No. 2, pp. 173–182, 1993.

[34] Hobbacher A., Recommendations for Fatigue Design of Welded Joints and Components, IIW document XIII-2151-07 / XV-1254-07, International Institute of Welding - IIW / IIS, 2007.

[35] Maddox S.J., Fatigue of Welded Joints, Seminar on Fatigue Performance of Weldments, The Welding Institute, Cambridge, England, 1983.

[36] Samuelsoon J., Integrated Design and Manufacturing of Welded Structures, Nordic Innovation Center, Norway, 2007.

[37] Nguyen N.T. and Wahab M.A., The Effect of Residual Stresses and Weld Geometry on the Improvement of Fatigue Life, Journal of Materials Processing Technology, Vol. 48, pp. 581–588, 1995.

[38] Singh P.J., Guha B. and Achar D.R.G., Fatigue Life Improvement of AISI 304L Cruciform Welded Joints by Cryogenic Treatment, Engineering Failure Analysis, Vol. 10, No. 1, pp 1–12, 2003.

[39] Hou C.-Y. and Charng J.-J., Models for the Estimation of Weldemnt Fatigue Crack Initiation Life, International Journal of Fatigue, Vol. 19, No. 7, pp. 537–541, 1997.

[40] Stoyan S., Kujawski D. and Mallory J., Fatigue Crack Growth in 2324 Aluminum Alloy, Western Michigan University, Technical Report No. MAE-05-01, 2005.

[41] Guo K., Bell R. and Wang X., The Stress Intensity Factor Solutions for Edge Cracks in a Padded Plate Geometry under General Loading Conditions, International Journal of Fatigue, Vol. 29, pp. 481–488, 2007.

[42] Bäckström M., Multi Axial Fatigue Life Assessment of Welds Based on Nominal and Hot Spot Stress, Ph.D. Thesis, Lappeenranta University of Technology, Lappeenranta, Finland, 2003.

[43] BEASY, <http://www.beasy.com>.

[44] Fricke W., Doerk O. and Grünitz L., Fatigue Strength Investigation and Assessment of Fillet Welds around Stiffener and Bracket Toes, Proceedings of OMAE specialty symposium on FPSO integrity, Houston, 2004.

[45] Fricke W. and Doerk O., Simplified Approach to Fatigue Strength Assessment of Fillet-Welded Attachment Ends, International Journal of Fatigue, Vol. 28, No. 2, pp. 141–150, 2006.

[46] Cornell Fracture Group, FRANC2D Version 3.2, <http://www.cfg.cornell.edu>.

[47] Hee S.C. and Jefferson A.D., Two-Dimensional Analysis of a Gravity Dam Using the Program Franc2D, Cardiff University, U.K., 2005.

[48] Ferrica J.A. and Branco C.M., Fatigue Analysis and Prediction in Filled Welded Joints in the Low Thickness Range, Fatigue and Fracture Engineering Materials and Structures, Vol. 13, No. 3, pp. 201–212, 1990.

[49] Kainuma S. and Mori T., A Fatigue Strength Evaluation Method for Load-Carrying Fillet Welded Cruciform Joints, International Journal of Fatigue, Vol. 28, No. 8, pp. 864–872, 2006.

[50] Nguyen N.T. and Wahab M.A., A Theoretical Study of the Effect of Weld Geometry Parameters on Fatigue Crack Propagation Life, Journal of Engineering Fracture Mechanics, Vol. 51, No. I, pp. 1–18, 1995.

[51] Ohta M.T. and Suzuki N., Evaluation of Effect of Plate Thickness on Fatigue Strength of Butt Welded Joints by a Test Maintaining Maximum Stress at Yield Strength, Engineering Fracture Mechanics, Vol. 37, No. 5, pp. 987–993, 1990.

[52] Bimalendu G., Effect of Specimen Geometry on Fatigue of Welded Joints, Engineering Fracture Mechanics, Vol. 46, No. 1, pp. 35–39, 1993.

[53] Niu X. and Glinka G., The Weld Profile Effect on Stress Intensity Factors in Weldments, International Journal of Fracture, Vol. 35, No.1, pp. 3–20, 1987.

[54] Motarjemi A.K., Kokabi A.H. and Burdekin F.M., Comparison of Fatigue Life for T and Cruciform Welded Joints with Different Combinations of Geometrical Parameters, Engineering Fracture Mechanics, Vol. 67, No. 4, pp. 311–326, 2000.

[55] Taylor D., Barrett N. and Lucano G., Some New Method for Prediction Fatigue in Welded Joint, International Journal of Fatigue, Vol. 24, No. 5, pp. 509–518, 2002.

[56] National Institute of Materials Science (NIMS), Data Sheet on Fatigue Properties of Non-Load-Carrying of Cruciform Welded Joints of SM490B Rolled Steel for Welded Structure-Effect of Plate Thickness (Part 3, Thickness 80 mm), Fatigue Data Sheet, No. 108, Japan, 2009.

[57] Radhakrishnan V.M., Welding Technology and Design, 2nd edition, New Age International Publishers, 2007.

[58] Itoh Y.Z., Suruga S. and Kashiwaya H., Prediction of Fatigue Crack Growth Rate in Welding Residual Stress Field, Engineering Fracture Mechanics, Vol. 33, No. 3, pp. 397–407, 1989.

[59] Barsoum Z., Residual Stress Analysis and Fatigue Assessment of Welded Steel Structures, Ph.D. Thesis, KTH Engineering Science, Sweden, 2008.

[60] O'Dowd N.P., Nikbin K.M., Lee H.-Y., Wimpory R. C. and Biglari F.R., Stress Intensity Factors due to Residual Stresses in T-Plate Welds, Journal of Pressure Vessel and Technology, Transactions of the ASME, Vol. 126, pp. 432–437, 2004.

[61] Lee H.-Y., Biglari F.R., Wimpory R. and Nikbin K.M., Treatment of Residual Stress in Failure Assessment Procedure, Engineering Fracture Mechanics, Vol. 73, No. 13, pp. 1755–1771, 2006.

[62] Assis J.T., Monin V., Teodosio J.R. and Gurova T., X-Ray Analysis of Residual Stress Distribution in Weld Region, JCPDS-International Centre for Diffraction Data, Advances in X-ray Analysis, Vol. 45, 2002.

[63] Shen W.Y. and Clayton P., Fatigue of Fillet Welded ASIS Steel, Engineering Fracture Mechanics, Vol. 53, No. 6, pp. 1007–1016, 1996.

[64] Cordiano H.V., Effect of Residual Stress on Low Cycle Fatigue Life of Large Scale Weldment in High Strength Steel, Transactions of the ASME, Journal of Engineering for Industry, pp. 86–92, 1970.

[65] Servetti G. and Zhang X., Predicting Fatigue Crack Growth Rate in a Welded Butt Joint: The Role of Effective R Ratio in Accounting for Residual

Stress Effect, Engineering Fracture Mechanics, Vol. 76, No. 11, pp. 1589–1602, 2009.

[66] Stacey A. and Webster G.A., Stress Intensity Factors Caused by Residual Stress Fields in Autofrettaged Tubing, ASTM, STP 1004, 1988.

[67] Maksymowicz A.Z., Magdon M., Whiting J.S.S., B. J.H., Chaufer B., Rabiller-Baudry M., Guihard L., Daufin G., Nguyen N.T. and Wahab M.A., The Effect of Undercut and Residual Stresses on Fatigue Behaviour of Misaligned Butt Joints, Engineering Fracture Mechanics, Vol. 55, No. 3, pp. 453–469, 1996.

[68] Nguyen N.T. and Wahab M.A., The Effect of Weld Geometry and Residual Stresses on the Fatigue of Welded Joints under Combined Loading, Journal of Materials Processing Technology, Vol. 77, No. 1, pp. 201–208, 1998.

[69] Barthelemy J.Y., Structural Integrity Assessment Procedures for European Industry- SINTAP, Compendium of residual stress profile, Final report BRITE-EURAM SINTAP, BE95-1426 Task 4, Institut de Soudure, 1999.

[70] Lee H.Y., Nikbin K.M. and O'Dowd N.P., A Generic Approach for a Linear Elastic Fracture Mechanics Analysis of Components Containing Residual Stress, International Journal of Pressure Vessels and Piping, Vol. 82, No. 10, pp. 797–806, 2005.

[71] Lee H., Lee S., Lee J., Wimpory R. and Nikbin K.M., Residual Stress Distributions for Plate T-Butt Welds in Defect Assessment Applications, Journal of ASTM International (JAI), Vol. 4, No. 1, pp. 1-10, 2007.

[72] Anderson T.L., Fracture Mechanics: Fundamentals and Applications, 3rd edition, Taylor and Francis Press, 2005.

[73] Kiciak A., Glinka G. and Burns D.J., Calculation of Stress Intensity Factors and Crack Opening Displacements for Cracks Subjected to Complex Stress Fields, Journal of Pressure Vessel Technology, Vol. 125, No. 3, pp. 260–266, 2003.

[74] Zheng X.J., Glinka G. and Dubey R.N., Stress Intensity Factors and Weight Functions for a Corner Crack in a Finite Thickness Plate, Engineering Fracture Mechanics, Vol. 54, No. 1, pp. 49–61, 1996.

[75] BS 5400 Pt. 10, British Standards Institution, 1982.

[76] BS 7608, British Standards Institution, 1993.

[77] Nykänen T., Marquis G. and Björk T., Fatigue Analysis of Non-Load-Carrying Fillet Welded Cruciform Joints, Engineering Fracture Mechanics, Vol. 74, No. 3, pp. 399–415, 2007.

[78] Wawrzynek P. and Ingraffea A., FRANC2D: A Two-Dimensional Crack Propagation Simulator, Version 2.7 User's Guide, NASA Contractor Report 4572, 1994.

[79] Iesulauro E., FRANC2D/L: A Crack Propagation Simulator for Plane Layered Structures, Version 1.5, user's guide, Cornell University, Cornell Fracture Group, Ithaca, New York.

[80] Balasubramanian V. and Guha B., Establishing Criteria for Root and Toe Cracking of Load Carrying Cruciform Joints of Pressure Vessel Grade Steel, Engineering Failure Analysis, Vol. 11, No. 6, pp. 967–974, 2004.

[81] Mashiri F.R, Zhao X.L. and Grundy P., Crack Propagation Analysis of Welded Thin-Walled Cruciform Joint Using Boundary Element Method, Materials Research Forum 1997 Conference Proceedings, Institute of Metals and Materials Australasia Ltd, Australia, November 20–21, 1997.

[82] Shukla A., Practical Fracture Mechanics in Design, 2nd edition, Taylor and Francis, Group, LLC, 2008.

[83] Erdogan F. and Sih G.C., ASME J. Basic Eng., Vol. 85, pp. 519–527, 1963.

[84] Singh P.J., Achar D.R.G., Guha B. and Nordberg H., Fatigue Life Predication of Gas Tungsten Arc Welded AISI 304L Cruciform Joints with Different LOP Sizes, International Journal of Fatigue, Vol. 25, No. 1, pp. 1–7, 2003.

[85] BSI7910, PD 6493 British Standard Institution, London, UK, Appendix J: 29, 1997.

[86] Lados D.A., Apelian D. and Donald J.K., Fatigue Crack Growth Mechanics at the Microstructures Scale in Al-Si-Mg Cast Alloys: Mechanics in the Near–Threshold Region. Acta Materilia Journal, Vol. 54, pp. 1475–1486, 2006.

[87] Brown W.F. and Srawley J.E., ASTM STP 410, ASTM, West Conshohocken, PA, 1966.

[88] Lie S.T., The Influence of Geometrical Parameters on the Fatigue Strength of Fillet Welds using Boundary Element and Fracture Mechanics Methods, Ph.D. Thesis, University of Manchester Institute of Science and Technology (UMIST), 1983.

[89] Thurlbeck S.D., A Fracture Mechanics Based Methodology for the Assessment of Weld Toe Cracks in Tubular Offshore Joints, Ph.D. Thesis, University of Manchester, Institute of Science and Technology (UMIST), UK, 1991.

[90] Bowness D. and Lee M.M.K., Stress Intensity Factor Solutions for Semi-Elliptical Weld-Toe Cracks in T-Butt Geometries, Fatigue and Fracture of Engineering Materials and Structures, Vol. 19, No. 6, pp. 787–797, 1996.

[91] Broek D., Some Considerations on Slow Crack Growth, International Journal of Fracture, Vol. 4, No. 1, 1968.

[92] Kainuma S. and Kimb I.-T., Fatigue Strength Evaluation of Load-Carrying Cruciform Fillet-Welded Joints Made with Mild Steel Plates of Different Thickness, International Journal of Fatigue, Vol. 27, No. 7, pp. 810–816, 2005.

[93] Örjasäter O., Effect of Plate Thickness on Fatigue of Welded Components SINTEF, Materials Technology Trondheim, Norway, IIW – XIII-1582-95, 1995.

[94] Yokota H. and Anami K., Local Stress Approach for Fatigue Assessment of Welded Joint, Kochi University of Technology, Japan, 2006.

[95] Liljedahl C.D.M., Brouard J., Zanellato O., Lin J., Tan M.L., Ganguly S., Irving P.E., Fitzpatrick M.E., Zhang X. and Edwards L, Weld Residual Stress Effects on Fatigue Crack Growth Behaviour of Aluminium Alloy 2024-T351, International Journal of Fatigue, Vol. 31, No. 6, pp. 1081-1088, 2009.

[96] Liu A.F., Structural Life Assessment Methods, 1st edition, ASM International, 1988.

[97] AFGROW, Fracture Mechanics and Fatigue Crack Growth Analysis Software Tool, <http://www.afgrow.net/>.

[98] Shehu M., Hübner P. and Cukalla M., The Behaviour of Fatigue-Crack Growth in Shipbuilding Steel Using the Esacrack Approach, Materiali in Tehnologije, Vol. 40, No. 5, pp. 207–210, 2006.

[99] Okamoto A. and Nakamura H., The Influence of Residual Stress on Fatigue Cracking, Journal of Pressure Vessel Technology, Vol. 112, No. 3, pp. 199–203, 1990.

[100] Almukhtar A., Hübner P., Henkel S. and Biermann H., Fatigue Life Calculation of Welded Joints with Fracture Mechanics Methods, DVM, Wuppertal, Germany, February 17–18, pp. 63–72, 2009, (In German).

[101] Lawrence F.V., Overview of Weldment Fatigue Life Prediction, University of Illinois, Urbana, USA, 1985.

[102] Callister W.D.Jr., Materials Science and Engineering: An Introduction, 7th edition, Willy & Son, 2007.

[103] Blake A., Practical Fracture Mechanics in Design, Marcel Dekker Inc. Publisher, 1996.

[104] Maddox S.J., Fatigue Strength of Welded Structures, 2nd edition, Abington Publishing, Cambridge, UK, 1991.

[105] Radaj D., Design and Analysis of Fatigue Resistant Welded Structures, 1st edition, Abington Publishing, 1990.

[106] Allen C.M., Gerritsen C.H.J., Zhang Y. and Mawella J., Hybrid Laser-MAG Welding Procedures and Weld Properties in 4 mm, 6 mm and 8 mm Thickness C-Mn Steels, The IIW Commission IV / XII, Intermediate Meeting, Vigo, Spain, April 11–13, 2007.

[107] R6, Revision 4, Assessment of the Integrity of Structures Containing Defects, British Energy Generation Ltd (BEGL), Barnwood, Gloucester, UK, 2001.

[108] MatWeb, Online materials information resource, <http://www.matweb.com/>.

[109] Sobczyk K. and Trebicki J., Fatigue Crack Growth in Random Residual Stresses, International Journal of Fatigue, Vol. 26, No. 11, pp. 1179–1187, 2004.

[110] LaRue J.E. and Daniewicz S.R., Predicting the Effect of Residual Stress on Fatigue Crack Growth, International Journal of Fatigue, Vol. 29, No. 3, pp. 508–515, 2007.

[111] Al-Mukhtar A., Biermann H., Hübner P. and Henkel S., Lebensdauerberechnung von Schweissverbindungen mit bruchmechanischen Methoden (Fatigue Life Calculation of Welded Joints with Fracture Mechanics Methods), 41 Tagung des DVMArbeitskreises Bruchvorgänge, Wuppertal, Germany, pp. 63–72, February 17–18, 2009, (In German).

[112] Al-Mukhtar A., Biermann H., Hübner P. and Henkel S., Fatigue Life Prediction of Fillet Welded Cruciform Joints Based on Fracture Mechanics Method, Proceedings of 2nd International Conference on Fatigue and Fracture in the Infrastructure, Philadelphia, USA, July 26–29, 2009.

[113] Al-Mukhtar A., Biermann H., Hübner P. and Henkel S., Fatigue Crack Propagation Life Calculation in Welded Joints, CD of International Conference on Crack Paths (CP 2009), Italy, pp. 391–397, September 23–25, 2009.

[114] Al-Mukhtar A., Biermann H., Hübner P. and Henkel S., Determination of Some Parameters for Fatigue Life in Welded Joints Using Fracture

Mechanics Method, Journal of Materials Engineering and Performance (JMEP), ASM International, Vol. 19, No. 9, pp. 1225-1234, 2010.

[115] Al-Mukhtar A., Biermann H., Hübner P. and Henkel S.,The Effect of Weld Profile and Geometries of Butt Weld Joints on Fatigue Life Under Cyclic Tensile Loading, Journal of Materials Engineering and Performance (JMEP), ASM International, DOI: 10.1007/s11665-010-9775-1, 2010.

I want morebooks!

Buy your books fast and straightforward online - at one of world's fastest growing online book stores! Environmentally sound due to Print-on-Demand technologies.

Buy your books online at
www.morebooks.shop

Kaufen Sie Ihre Bücher schnell und unkompliziert online – auf einer der am schnellsten wachsenden Buchhandelsplattformen weltweit! Dank Print-On-Demand umwelt- und ressourcenschonend produziert.

Bücher schneller online kaufen
www.morebooks.shop

KS OmniScriptum Publishing
Brivibas gatve 197
LV-1039 Riga, Latvia
Telefax: +371 686 204 55

info@omniscriptum.com
www.omniscriptum.com

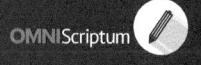

Printed by Books on Demand GmbH, Norderstedt / Germany